PENGUIN BOOKS

SIEGES OF THE MIDDLE AGES

Philip Warner, a native of Warwickshire, where his family have lived since 1600, and a graduate of Cambridge University, is the author of forty-eight books, mainly on military history, including four on castles, all of which required extensive travel and research.

He served in the army throughout the Second World War, mainly in the Far East, and was subsequently an Assistant Principal at HM Treasury, working in the Defence Materials Division, a Lecturer for the British Council in Spain, and a Senior Lecturer and Head of Communications Studies at the Royal Military Academy, Sandhurst. He has been a military obituaries correspondent for the *Daily Telegraph* since 1987. He is a member of the Royal Archaeological Institute. When not researching or writing, his recreations are rugby, football, fly-fishing, sociable wine-drinking, conversation and music of all varieties. He is a member of the Athenaeum, the Harlequin Rugby Club, Jester's Squash Club and Sandhurst's trout-fishing club.

SIEGES OF THE MIDDLE AGES

PHILIP WARNER

PENGUIN BOOKS

PENGUIN BOOKS

Published by the Penguin Group
Penguin Books Ltd, 27 Wrights Lane, London W8 5TZ, England
Penguin Putnam Inc., 375 Hudson Street, New York, New York 10014, USA
Penguin Books Australia Ltd, Ringwood, Victoria, Australia
Penguin Books Canada Ltd, 10 Alcorn Avenue, Toronto, Ontario, Canada M4V 3B2
Penguin Books (NZ) Ltd, Private Bag 102902, NSMC, Auckland, New Zealand

Penguin Books Ltd, Registered Offices: Harmondsworth, Middlesex, England

First published by G. Bell and Sons, Ltd 1968
Published as a Classic Penguin 2000
1 3 5 7 9 10 8 6 4 2

Printed in Great Britain by CPI UK

Contents

Plates

Diagrams and Maps

Preface

ANYONE who writes a book on castles soon finds himself owing a large debt of gratitude to many people. There are librarians who track down obscure and rare books, kind friends who take photographs, owners of land on which castles once stood, and people who make encouraging and helpful suggestions. Among so many it may seem unjust to single out names but I find it necessary and just to mention one or two. First, there is Brigadier Peter Young, D.S.O., M.C., who suggested I should write this book, and secondly, there is W. L. McElwee, M.C. who convinced me I could. Both have been extremely liberal with their encouragement and criticism. Few books can have had as much constructive and varied criticism as this for it has had to pass the scrutiny of my family who claim to represent the 'general reader'. My daughter Diana in particular took every sentence and shook it to see if it would fall to pieces.

Professor Dorothy Whitelock very kindly gave me permission to quote from the *Anglo-Saxon Chronicle*, which she edited with Professor David Douglas and Miss Susie L. Tucker. Mr K. R. Potter very kindly gave me permission to quote from his translation of *Gesta Stephani*, and Mr J. T. Appleby from his translation of *Richard of Devizes*.

I am particularly grateful to Mrs Blanche Ellis who took enormous trouble over maps, diagrams, and illustrations; whenever possible she drew from the original weapon or piece of machinery.

My special thanks are due to Richard Warner who spent part of his holidays translating difficult mediaeval Latin texts, John Warner, who helped with research, and all those kind people who, hearing I was writing about castles, sent me pamphlets, cuttings, or photographs, in the hope that I would find them interesting, which I invariably did. Nothing was too much trouble for Colonel Alan Shepperd, M.B.E., and his splendid staff at the Royal Military Academy Sandhurst Central Library, and no one could have been more patient with my difficult requests than the Librarian and staff of the County Library at Camberley.

P. W.

❋ 1 ❋

Introduction

THE word 'castle' is charged with emotion. To some it represents a gallant survival of a romantic and chivalrous past, to others it is the symbol of an Englishman's pride and liberty (his home is his castle), for others it represents a golden age when everyone knew his place and kept to it. Wildly inaccurate though they be, the existence of such beliefs is not surprising. The castle appears to offer an easy entry into the past, it looks both romantic and independent, and it belongs to an age in which class barriers were approved and enforced.

Standing on the battlements of a castle the humblest person feels a sense of power and grandeur. He is back in the past and feels a kinship with the original owners. In all probability this kinship is genuine, though remote. Every family that was in England in 1087 is now related thirteen times over to every other family in the country at that time; he is thus related both to the mighty baron and the most downtrodden villein. But this thin tie of blood is the only link he has with an age that ceased to exist five hundred years ago.

It is almost as difficult for him to imagine that world as it would be for a twelfth-century knight to visualize a modern city. It is not just the way of life that is different, it is the whole process of thought. In studying any feature of the Middle Ages it is essential to keep this difference in mind.

The function of a castle was to provide a refuge, and dominate an area. It also served as a residence, storehouse, administrative headquarters, gaol, barracks, symbol of authority, and observation point. Castles had uses which varied according to the place they occupied, and the countryside they controlled. Some were for an attacking strategy, such as Henry II's in Ireland, others were for deep defence in remote Welsh valleys. They could be manned by small forces, yet in time of need would accommodate a large number of troops. In forward positions they could gain priceless time while the countryside to the rear was being prepared against an invader: if bypassed they could be a perpetual menace to enemy communications. They were one of the most useful devices ever invented but they had one great drawback; they were expensive and difficult to build, and once built they were always in need of costly adaptation or development. In the course of time many powerful castles have disappeared without trace; Reading, Newbury, and Farringdon are examples.

Surviving castles fall into two categories. Some have been modified for residential purposes, and surrounded by attractive gardens: Windsor and Warwick are of this type. Others, such as Dinas Bran (North Wales), and Lewes, are ruins and are too far gone to give a clear picture of what they were once like. Both types are so quiet and dignified the visitor hesitates to raise his voice.

But in their day castles were centres of noise and bustle. In peacetime the castle wards would be like a noisy market; in wartime they would be like factories, piled high with stores, and with a host of supporters backing the front-line defence. To-day the peaceful walls that crown a steep hill give an entirely false impression of the castle as a form of passive defence; a retreat in which one would be protected by difficulty of access. Access was indeed made as difficult as possible for the unwelcome, but the overriding thought in castle strategy was not passive defence but action and destruction. Shutting oneself up in a castle was not an attempt to avoid conflict, but a manœuvre to make the enemy fight at a disadvantage. Along the castle approaches would be chosen 'killing grounds' where its attackers would be exposed to fire without being able to return it effectively. Even an incompetent and cowardly commander would benefit by the lessons built into stone by his predecessors. The defence had an enormous advantage. To an invader time would be vital, and it would be important to maintain the full strength of his army lest he should be outnumbered on the battlefield. Detaching small forces for sieges would ultimately leave him numerically inferior. The enemy might be an invading army, anxious to press on but unwilling to leave an uncaptured fortress on its line of retreat. The castle would have to be besieged, and perhaps taken, but the designer, who had probably also chosen the site, would have tried to ensure that the siege would be costly in time and lives. It would not always be so, for fixed defences often defy careful military calculation. With few exceptions, such as Kenilworth and Harlech, castles did not stand long sieges; starvation saw to that. But besiegers had their own problems which sometimes became so pressing that a siege was abandoned. They were exposed to the weather, they lost men through desertion, and might be shorter of food than the people they were besieging, for the latter would have cleared the countryside of supplies before pulling up the drawbridge. They might even be besieged themselves, as happened at Wallingford in 1152. Furthermore, they might be given a thoroughly unpleasant time by those they were trying to besiege. Froissart describes the siege of Aiguillon in 1346 when the English were surrounded by a large host of French.

The French battered them with missiles from twelve engines 'but, they within were so well pavised (protected) that never a stone of their engines did them hurt. They within also had great engines, the which brake down all the engines without, for in a short space they brake all to pieces the greatest of them without.'

Not content with mere counter-fire, there were frequent sallies of a hundred or so men. As these were intent on bringing in supplies they were usually engaged by the French at some point in their foraging. On the majority of these occasions the attackers received the worst of the encounter.

But, if life was difficult for the besieger outside the castle perimeter, it was doubly so once he came closer. The moat might be wide and deep, and contain sharpened stakes; a hail of missiles would rain down on him from the battlements, and if he broke through the walls he might well find himself in a trap. Finally, within the inner ward, he would have to fight his way up steep winding stairways where every advantage was conferred on the defenders. Yet, in spite of all these hazards, there was never an impregnable castle. Not all castles were captured, because some were never attacked, but the lesson of history was that no man can make a defence that other men cannot break through. Château Gaillard, the brilliant construction of Richard I, was thought to be impregnable, but what happened to it is described later in this book. The best that a besieged castle could hope for was to raise the price of victory to a point which the besiegers would be unwilling to pay, a price not only of men and time but also of siege materials, which might have to be brought a hundred miles or more. Some of the equipment used in the siege of Rochester came from the Forest of Dean. Caerphilly, a masterpiece of castle design, was a 'high-price' siege, and was left alone after 1327.

Steep slopes and isolated positions tend to be associated with castles nowadays because that is where ruins have survived. But many strong castles were built on flat ground or gentle slopes. Shirburn, Wallingford, and Boarstall are examples; the last of these is in a hollow and probably blocked the main trackway across a marsh. It does not look very formidable to-day but in its heyday the moat was 60 feet wide and the other defences in proportion. Such castles relied on strong walls, or large garrisons, or marshy approaches, or wide moats, for their ability to disconcert the attacker. They enjoyed several advantages over their more lofty counterparts. They were more comfortable residences, they could not quickly be starved into submission; and supplies were more easily brought in during peace. But however attractive these amenities the invader of mountainous country like North Wales would have to forgo them, or he would himself be assailed from nearby peaks.

Anyone who writes about castles relies heavily on the work done by architects and archaeologists who have elucidated and explained features that might have been misunderstood or neglected. The fact that some of their deductions have been disproved does not make their technical descriptions less valuable and many of their theories are at least as tenable as those of their historian critics. But the castle can only be appreciated if it is seen from its beginning to its decline in the military and political setting that caused its rise and decay. In that process it served many different purposes.

The English castle, as we know it, has French origins. It first appears in this country before the Conquest, when in Herefordshire, and perhaps elsewhere, a few Normans, invited over by Edward the Confessor, built mound castles and attracted the hatred of the local people. Richard's Castle, built by Richard Fitz Scrob, near Ludlow, had a motte 70 feet high, with a flattened top 30 feet in diameter. Around it was a deep ditch; outside this was a palisade and then a smaller ditch. This type of defensive/offensive structure had been developed in France nearly two centuries before, and differed from fortifications in this country in that it was held by a single owner, who in turn held it from the king. It symbolized the feudal structure of the state as a pyramid with the king at the top. Castles were built and held under royal licence. Their owners were tenants of the king; they in their turn had tenants owing allegiance to them. At the highest level the 'rents' were not particularly onerous; Weston, in Warwickshire, was held for a brach (hound) and 7d each year.

The Norman concept of defence was vastly different from what had preceded it in this country, but was not, at first, an advance. Prehistoric earthworks, as found at Maiden Castle, Dorset, Cissbury in Sussex, Old Sarum, Wiltshire, and Blackbury Castle, Devon, show ingenious arrangements of diversions in which an attacker could be trapped and exterminated. Some of these 'earthworks' (occasionally built of stone) incorporated military sophistications that were not seen again until English castle building reached the height of its achievement in the fourteenth century. In that period English designers skilfully incorporated the lessons of 4000 years, and built fortifications superior to any in the world.

Roman defences were walled towns or camps, with ramparts and ditches. In the heyday of Roman power towns were usually square (Pevensey was an exception) and four straight streets led from each gate to the centre (Plate 6). A Roman town was merely a base for a highly trained body of men who could traverse the country rapidly over the excellent roads they had made. This was mobile defence by the strategic reserve.

Defences built between the departure of the Romans and the arrival of the Normans are usually loosely classed as 'burhs'. These were townships protected by an earth bank and a stockade. The stockade was usually a formidable obstacle but their main strength appears to have lain in the difficulty of access. Marshy and treacherous land was plentiful, and full advantage was taken of it.

The skill shown in choosing and siting early fortifications, whether prehistoric, Roman, Saxon, or Norman, was not fully appreciated till 1940 when Britain faced the threat of invasion. In that year the country was carefully surveyed for active defence; every site of strategic value was found to bear traces of previous military use.

This book is concerned with the evolution of siege warfare from the arrival of the Norman castle in England in 1066 to its decline in the fifteenth century. The end of our period is not, of course, the end of the castle for it played a formidable part in the Civil War two centuries after its day was supposed to have been over. However, by the beginning of the fifteenth century the patterns of warfare had changed. Issues were decided by battles in the open field, and this process culminated in the crown of England changing hands at Bosworth in 1485, an occasion when some 20,000 men disputed the future of the kingdom. The castle ceased to be an instrument of warfare and, being inconvenient and uncomfortable as a residence, was soon drastically modified, or abandoned altogether.

Although many of the sieges described took place overseas they are English sieges by the fact that they were the concern of armies from this country. Siege warfare has a long and interesting history and some of its techniques were well developed as long ago as 3000 B.C. The lessons learnt were embodied in the fortifications of the Eastern Mediterranean, and these were absorbed by travellers from Europe, of whom the most famous was Richard I, in the mediaeval period.

Although the type of siege warfare described in this book belongs to the past, the concept of siege is still with us. Nowadays, however, it is on a vast scale. The Berlin Wall is the visual symbol of the Iron Curtain which divides Fortress East from Fortress West. Viet Nam has been in a state of siege for many years. Many of the battles of the First World War, such as Ypres and Vimy Ridge, were forms of siege, and the Second World War saw the investment of towns, peninsulas, and islands. Britain herself was under siege between 1940 and 1944, attacked through submarine warfare and aerial bombardment. Malta lasted out, Crete was overwhelmed, Singapore and Hong-Kong were doomed as soon as they had lost their highly vulnerable water-supply. Tobruk was another notable siege, and Stalingrad undoubtedly saw some of the closest and bitterest fighting of the war. Iwo Jima and Okinawa saw brief but intense sieges, but if the invasion of Japan itself had taken place this might well have ranked as the bloodiest battle in history.

Although these recent sieges are remote in time, and differ vastly in the weapons and materials involved, one does not need to be a military historian to see striking similarities of principle and technique between them and their mediaeval counterparts.

P. W.
1968

* 2 *

The Development of Siege Warfare Techniques

THE siting of castles was governed by two factors: strategic necessity and an eye for ground. Strategic necessity dictated that castles had to be built at certain points, but the exact position was determined by the possibilities of the immediate area. 'Capability' Brown, the great eighteenth-century landscape gardener, earned his nickname for the remark he would make on surveying open countryside: 'This has capability.' Subsequently streams would be dammed, trees planted and soil scarped, until a home such as Blenheim Palace was framed in a perfect setting. The same technique was used in castle siting. Certain situations such as Wallingford and Oxford, guarding vital fords, selected themselves. The same would be true of Dover and Rochester, blocking the gateway from the Continent. But Kenilworth might have been sited by a military 'Capability' Brown; its value depended not on its natural strength but on the artificial barriers that were created by manipulating two small streams. Leeds Castle (Kent), Shirburn and Broughton (Oxon), and Caerphilly (Glamorgan) were created in a similar fashion (Plate 11). A favoured situation would also be a slope or spur which could be cut off from the rest of the ridge by a deep ditch: Château Gaillard is an example of this method. The mediaeval builder liked to work with nature rather than to defy it as often happens to-day.

Some castles owed their existence to the need for overawing a neighbouring township. Exeter, Winchester, York and Nottingham are of this variety. In Stephen's reign most of the adulterine castles were probably built to oppose a neighbour or dominate a district. When Henry II came to the throne he had most of these castles demolished, and where ruins do remain it is an insoluble puzzle to determine what they were defending and where the threat came from. Under the feudal system all castles were built, fortified and held under royal licence, but when the monarch was weak, as with Stephen, Henry III, Edward II, Richard II, and Henry VI, illegal fortification flourished.

1940 drew attention to many long-forgotten strategic and tactical positions because in that year Britain faced both internal and external threats. The external threat—invasion by sea—could very well have resulted in a second battle of Hastings, the internal threat by parachute and glider could have re-enacted many of the minor battles and skirmishes of our turbulent past. The fact that in some areas the Home Guard were partly equipped with bows and pikes would have added a macabre authentic flavour.

The term 'strategic factors' means hills, mountains, rivers, marshes, and vegetation. Needless to say, these vary considerably in importance and quality, but the presence of any of them must be significant to some degree. Before the Second World War much use was made of the term 'impenetrable jungle' but Burma and Malaya demonstrated that there are ways through or round almost every obstacle that nature can produce. Conversely a very simple feature can disturb an army and contribute to its defeat. Sedgemoor is an example. The best known instance of disaster occurring through miscalculation of natural obstacles was John's experience in crossing the Wash. He was caught between a high tide and a fast current on the Welland; he lost his equipment, his campaign, and ultimately his life.

Hilly or mountainous country is, of course, a formidable obstacle for any army, and therefore it is natural to find fortifications guarding the easier gaps. Skipton Castle (Yorks) controls the Aire gap—the principal crossing point of the Pennines. Reigate Castle (Surrey) was neatly situated to control the crossing of the east-west road with that of the north-south. Behind lay the passage through the North Downs. But in England mountains and hills did not have the importance they attained in other countries.

River and marsh were the major military obstacles of the middle ages. Accordingly, Berwick Castle controls the Tweed crossing, Newcastle and Corbridge the Tyne. An army moving south would find its next biggest obstacle in the Aire, and have to contend with Castleford and Pontefract. If it reached the Trent the alternatives would be the crossing controlled by Nottingham Castle, or a problem equally formidable at Newark.

To be worthy of the name a ford should be wide enough to allow an army to pass over fairly quickly. Narrow fords and slow crossings were likely to be fatal, as was proved on more than one occasion. Although many of these control points are commemorated in their names, some are not. Stamford controlled the Welland, but Huntingdon did the same for the Ouse. Rochester commanded the Medway; Cricklade, Oxford, Wallingford, Reading, and Windsor supervised the Thames.

A number of powerful castles adjoined rivers whether the latter were fordable or not. A river was not merely an obstacle to movement overland but also a means of advance or supply. The Trent was a water road for the Midlands in much the same way as the Thames served the south.

The explanation of castle siting may not always be obvious to-day. Rivers have been bridged, ports silted up—as Ravenspur at the mouth of the Humber—and ancient trackways abandoned for modern roads.

Draining of marshes and removal of forests have made communications easier, and therefore blurred much of the strategic significance of the old routes. Where forest has gone it has probably gone for ever but marshland can return with surprising ease. In 1917 the Allies decided to soften up the approach to Passchendaele with a heavy artillery bombardment. Although warned that this would upset the local drainage arrangements they persisted in their policy. As a result Passchendaele was fought in a sea of mud. But a return to such conditions does not need anything so dramatic as a major battle; all that is required is the neglect of dykes and drainage, which might result from freak weather conditions. In creating the dust bowl of the Middle West no one had any doubts that all was for the best—until nature reasserted herself. Modern British farming has the same happy self-assurance.

The four main rivers of Britain are the Thames, Humber, Trent, and Severn. Of these the Severn has figured least in strategy, but such names as Bristol, Gloucester, Worcester, and Shrewsbury, to name but a few, show that it has played its part. The presence of strong castles at Warwick, Northampton, Derby, Tamworth, and Farnham is explained by the need to establish bases at central points. It should be borne in mind that the influence of a castle was not bowshot range, but a comfortable day's ride; normally this amounted to about twenty miles radius from the castle but could be much more or much less according to the nature of the country. Border castles were centres of refuge. Callous and indifferent to human suffering though mediaeval barons were, they had a fine regard for the value of their own tenants. This was particularly so after the Black Death had thinned the ranks of agricultural labourers. As a result a feudal lord took considerable care to see that marauding neighbours did not slaughter his work-force.

Edward I's Welsh castles were not only examples of brilliant architecture, they were also extremely well-sited. Caernavon commanded the Menai Strait, Conway controlled the coast road from Chester to Snowdonia, and Beaumaris (Plate 14) protected Anglesey, which was the granary of Wales. All were sited so that they could be supplied from the sea.

Although castles were symbols of independence and individualism, there was more co-operation between their holders than is commonly supposed. In Wales mutual support was absolutely essential if the Normans were to survive. Fortunately for the invaders the native Welsh hated each other so much that there was seldom any concerted resistance to the Norman or English conquest.

In considering the essentially English character of the English castle it should not be forgotten that castle-building was introduced and developed by the Normans, that architects, supervisors, and even masons had to be imported from Normandy, that even the exterior stone was shipped over from as far away as Caen, that even as late as the fourteenth-century we were still drawing some of our ideas from France (as shown in Warwick castle). In these circumstances it is remarkable that English castles had an individuality that makes them easily distinguishable from their foreign counterparts, to which they were in no way inferior.

Although many castles were sturdy and forbidding structures

capable of withstanding heavy attack, these qualities were not always essential. Castle strategy was designed to blunt, divert, and dishearten an aggressor; it could decide a campaign without a pitched battle or too much bloodshed. This was a matter of mathematics. If one thousand men invade a country which has five castles set along its frontier, one hundred men may be given the task of reducing each point. The invader has thus reduced his force to 500 without achieving anything but a further liability, for his besiegers may be caught between a castle sortie and a local attack. Without a 10–1 superiority he is unlikely to achieve a quick victory over a castle; and as often as not the campaign ends in an ill-tempered withdrawal. The next year it all begins again.

The first castles built by the Normans in this country were Motte and Bailey structures. Motte was the Norman French word for turf; bailey means a palisade around an enclosure. The area inside the bailey was the guard, but as the Norman writers made no distinction between G and W the word often occurs as ward, and is used in this form at Windsor to-day. The motte was a heap of earth from ten to one hundred feet high, flattened at the top, and provided a platform that might be 300 feet across. On this platform was built a wooden tower, known as the donjon, a derivation from debased Latin meaning 'dominating point'. Originally it signified the highest point, but when later the tower was abandoned for more comfortable residential quarters, the donjon was used as a prison. When the prison sank from the highest to the lowest point in the castle it took the word 'dungeon' with it. While this process was going on, the word motte also descended, and was employed to describe the dry or wet ditch that surrounded the mound or bailey.

Mottes were easily and quickly erected. William I built one at Hastings immediately after he had arrived in this country. This motte appears on the Bayeux tapestry. The attacker had to negotiate a ditch full of water or sharp spikes, storm up a slope too steep for horses, dismantle a palisade of stakes and thorns, and finally capture a tower full of desperate men. In these trying circumstances his best ally would be fire, and this weapon was a most potent force long after wood had been replaced by stonework. Defence against fire was not easy and the best that most defenders could manage was a protective wall of damp hides. These were not particularly effective and had a disadvantage in that the attacker could use them to help him cross the spikes.

The tower (donjon) was a well-designed structure of two or three stories. Lambert d'Ardres gives an account of an elaborate one in France, but those in this country were, as far as we can judge from excavated post-holes, much simpler affairs. The tower erected at Hastings in 1066 had been made in Normandy, and consisted of jointed sections. Shipping it provided a major problem in logistics; the time, effort, and ingenuity used in arranging the transport shows the importance attached to it.

The motte had certain advantages over previous fortifications in this country. Although a residence, it was essentially a fighting device designed to work with maximum economy and efficiency.* In contrast the 'burh' had been a place of refuge for non-combatants, a centre of trade, and an un-unified command. The Normans were not concerned with defending towns or, in the early stages, anyone but themselves. Comfort was ignored. Cooking was done out of doors; the only solace for mental or physical ills was wine. However, as the experience of campaigns in this century has shown, life can be reasonably tolerable in conditions of dirt, discomfort, and cold, perhaps because hard living blunts desires, and constant danger makes a man value what he has, rather than pine for the impossible.

Within a short time the defences would be extended by the bailey, which in turn would have a defensive ditch. The area enclosed would be used for keeping cattle and other stores in times of emergency. The motte and bailey, seen in section, are shown in Figure 1.

Bridges were portable constructions that could be removed speedily in the event of impending attack.

As soon as possible, wooden structures were replaced by stone. As many years would have to pass before artificial mounds could carry a stone structure the Normans had to look for other sites for their stone keeps. Building in stone is a lengthy and expensive process so the mottes remained in use for many

* Recent excavations have shown that there was more variety in these castles than had previously been realised. Post holes show that strong gatehouses existed on some mottes. Thirteenth-century designers were not, therefore, quite the innovators they were often thought to be.

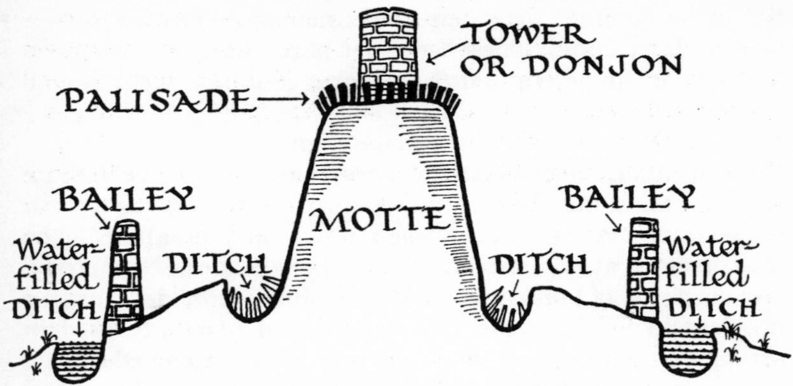

Figure 1. Section of Motte and Bailey.

years. In certain areas, where timber was scarce and stone plentiful the latter had been used from the first. But for the majority of the country stone buildings meant fresh sites. Hence one often finds the remains of a motte and bailey near a stone castle, or even enclosed in its inner perimeter.

A common compromise was to build what was known as a 'shell-keep'. This amounted to no more than replacing the wooden bailey with a formidable stone wall. The donjon would gradually revert to non-military uses while the main gateway would be strengthened and take over the main defensive function. Berkeley, Launceston, and Cardiff castles are all examples of this type of development (Plate 10).

The Normans had very little grasp of the principles of architecture and most of their buildings were stubbier and more solid than they needed to be, even allowing for bombardment with heavy missiles. The square keeps that they began building in the second half of the eleventh century were immensely strong having, in some castles, walls twenty feet thick. As may be seen in many ruined castles to-day walls were made of flints bound together by highly efficient cement, and often covered over with ashlar blocks. The keep at Dover has walls up to twenty four feet thick in places. Rochester's walls are only half that thickness but Rochester, as we see later, was breached by fire (Plates 7 and 8). The base of walls was usually built out, particularly at the corners. A splay at the base offered a greater obstacle to the picks of the attackers, and had an additional advantage that missiles dropped from above bounced and ricocheted into the assailants who were bound to be partly exposed at the sides. The grim aspect of these Norman keeps is well exemplified by Dover in Kent, Conisborough in Yorkshire, and Kenilworth in Warwickshire.

However it rapidly became clear that more than a forbidding appearance was needed to deter would-be attackers. The square keep had vulnerable points at the four corners. In the early castles defence was conducted, for the most part, from the battlements. In consequence the attack concentrated on the corners which were difficult to defend from an angle, but comparatively easily and rewarding to hack into. A wide breach in a corner would affect two walls instead of one and might bring down a large part of the structure. The answer to this was to build towers on the corners and, if necessary, along the side walls. This principle was soon used on the bailey also, and the castle began to resemble Figure 2.

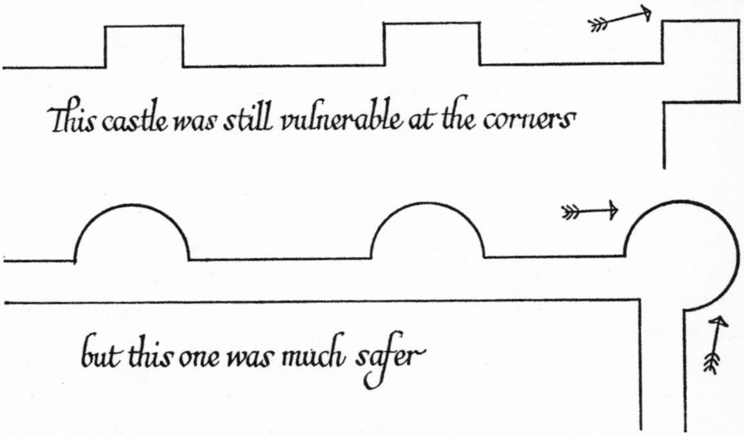

Figure 2. The development of Bastions.

These protruding towers gave covering fire along the walls and protected each other. Some of the walls and towers, such as Guy's and Caesar's at Warwick (Plate 15), seem unnecessarily high to modern eyes but there were sternly practical reasons for these lofty structures. Scaling ladders had their limits and launching machines lost much of their effectiveness if they were forced to attempt high trajectories. Stones dropped from a considerable height arrived with an impact sufficient to smash through mantlets, bores, and battering-rams.

Crenellations varied, but were seldom equally divided between the merlon (stone part) and embrasure (the gap). Merlons were usually about five feet wide and seven feet high. The embrasure also had a protective wall rising about three feet from the level of the rampart walk. Additional defences were roofing slopes to deflect missiles from above and shutters in the embrasures. Shutters were like hanging doors and were sometimes in two sections.

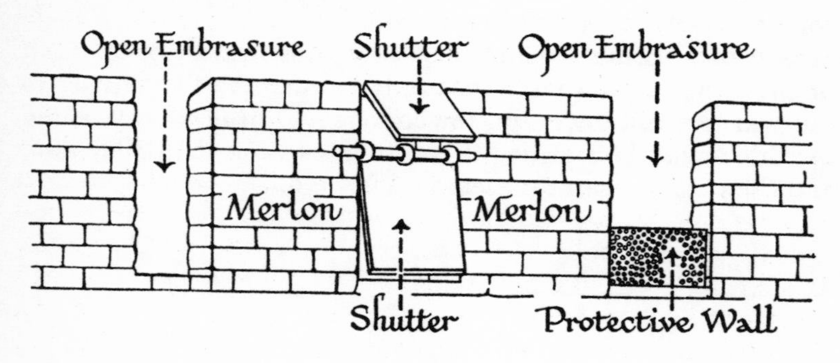

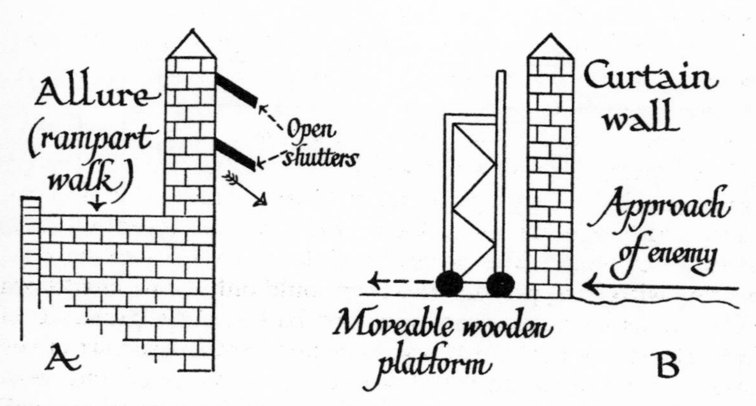

Figure 3. Battlements, Allure, and Curtain Wall. Note the very low parapet on the inside of the allure (or rampart walk). If an attacker gained the allure, he would be exposed to fire from the interior defences.

The rampart walk was called the allure. It had no protection on the inside so that if a wall was scaled the attacker would be fully exposed to fire from the rest of the castle. The disadvantage of battlements lay in the difficulty of attacking people immediately below. Accordingly, shelves were built out from the parapets and given the name 'hoardings' or brattices; later, when built with stone they were called 'machicolations'. As far as is known they were first used in this country in 1187, at Norwich, but they are known to have been used in the Middle East at least 1000 years B.C. Hoardings were very useful in defence because they could easily be erected or moved. On towers or walls they projected about four feet, and had slots in their floors. Popular belief held that boiling oil was poured down on the assailants but this is highly unlikely owing to the cost and difficulty of supply. A more likely substance was water, hot or cold, which might put out fires started at the base of the walls and prove inconvenient to those tending them. The wall along the bailey, sometimes described as the curtain wall—was usually protected by hoardings or platforms on the inside. Ideally this wall should be both high and thick so that anyone who managed to scale it would promptly be picked off by defensive fire as soon as he appeared on the top, or attempted to slide to the ground inside. An allure would be a dangerous provision on a curtain wall for it might prove more useful to the attacker than the attacked. Where allures did run along the curtain they were usually partitioned off at intervals. This was designed on the same principle as a ship's watertight bulkhead, to prevent damage spreading.

Internal passages had portcullises. These were gratings or grilles which were lowered from slots in the roofs. Warwick castle provides a good example of blocked internal communication, although this must have been tiresomely inconvenient in times of peace. The gateway of the barbican had no connection with its upper stories which could only be reached from battlements or a staircase from the bailey. This principle of water-tight compartments was to ensure that if any part of the structure was breached the attack could be contained and, with luck, destroyed. An attacker who had broken in might well find further advance impossible and his retreat cut off. The portcullis was designed to create such situations but it was neither as successful nor as widely used as might have been expected. In theory this great iron gate or grill was dropped behind assault troops who had broken through the outer gate. They could then be trapped in a confined space and effectively dealt with through slots provided for the purpose, and known as 'meurtrières' (murder holes) or killed by determined men emerging from side passages. In practice several mishaps could occur. The portcullis could jam of its own accord, or more probably would be held up by baulks of timber specially brought in for the purpose. Even if the portcullis did trap the

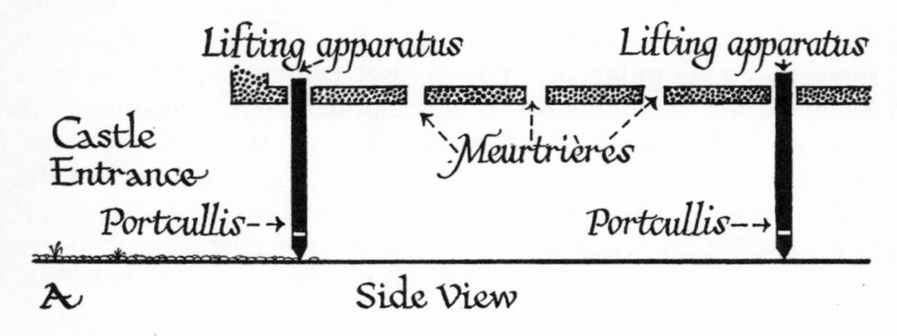

THE PORTCULLIS

Figure 4. The Portcullis. *A.* Side view illustrating the defensive plan of a gateway with two portcullises. *B.* Front view of the portcullis when lowered. *C.* The lifting apparatus for the portcullis of the Bloody Tower in H.M. Tower of London: it is still in working order.

front wave of the attackers all might not go smoothly for the defence, for fighting in a confined space is an unpredictable activity, and it might not be clear who was actually at a disadvantage. Still it must be noted that on several occasions the portcullis device proved extremely effective.

As mentioned above, in the earlier stages defence was conducted from the battlements. However, towards 1200, arrow loops began to appear lower down and, later still, they appeared in the merlons. The narrowness of most loops is a tribute to the accuracy of archers who, themselves under fire, could put a bolt through a slender aperture from a distance of thirty or forty yards. Still, as we know from the William Tell legend, this quality of marksmanship was not uncommon. Arrow-loops might be as narrow as half an inch wide, though most were of two or three inches. They were up to seven feet high and widened inwards at an angle of about 45°. This angle gave defenders a reasonable field of fire but must have been extremely cramping. They were usually manned by two defenders so that fire could be reasonably continuous. Later castles

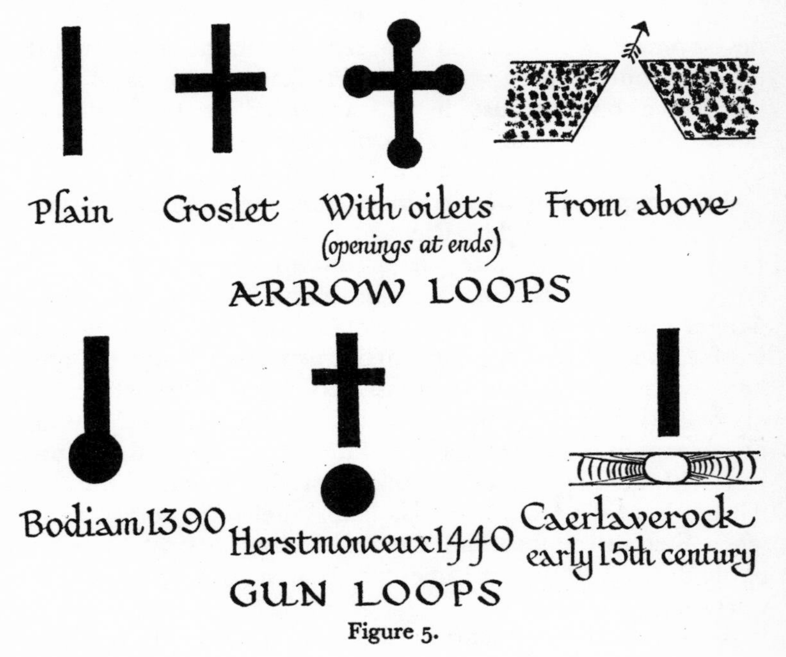

Figure 5.

had horizontal slots which effectively widened the range (Figure 5).

With the introduction of gunpowder, loops were widened considerably and are known as gun-ports. The over-caution noticeable in the design of arrow loops indicates that at all stages conservatism concentrated defence on to the battlements. Curiously enough, windows, which were at first used for ventilation rather than lighting, were much more vulnerable than arrow loops. This was brought out in the siege of Château Gaillard. In theory a window could always be shuttered in times of emergency; in practice, through neglect or treachery, this was often omitted.

Later, when windows became larger they were secured with strong iron grilles. Many of these have now rusted away, and their absence makes the apertures look extremely vulnerable; in their day, however, these grille-protected windows were very secure.

The absence of arrow loops in early castles was mainly due to the low esteem in which archery was held. This was not surprising as the early bows lacked power and range. The development of the crossbow caused a dramatic revision of this opinion, but archery was now criticized not because it was ineffective but because it was regarded as too cruel and devastating a weapon. The crossbow was, in fact, banned by the Lateran Council of 1139, as being too inhuman, but its unpopularity probably derived more from the fact that it was too effective against knights. Richard I was killed by an arrow in 1199. Mediaeval warfare was organized to provide adventure and entertainment for the aristocracy without much more danger than hunting or the tournament. The crossbow made warfare impersonal, perilous, and frustrating. What the crossbow began the longbow completed, as the French knights discovered in the slaughter of Crecy. Underrating and opposing new weapons is a characteristic of military conservatism. It was seen with the machine-gun, tank, aircraft, and parachute drop. All of these were criticized as being useless or impracticable when first introduced but the real feeling against them probably derived from a sentimental attachment to unmechanical warfare.

By the thirteenth century the inadequacies of the square

keep had been exposed on numerous occasions, and it was gradually being replaced by polygonal or octagonal structures. These were sound tactically but made inconvenient accommodation owing to their curved walls and projections. The keep was, however, due for even more drastic changes. The old concept of an inner citadel, whether square, or shell-keep such as may be seen at Windsor, was about to be abandoned. Experience had shown that once the bailey was breached the keep must inevitably fall, although there might be some delay before that occurred. An inner citadel was a semi-passive form of defence and, as pointed out earlier, the technique of castle tactics was offensive defence.

Accordingly the keep disappears, and is replaced by the heavily fortified gatehouse. This being at the main entrance, and theoretically the weakest point in the structure, was the most likely place for an attack to occur. By the same token it would offer the best opportunity for the defence to destroy the assailant. The approach road would have most of the characteristics of a first-class ambush, with frontal fire, cross fire, diagonal fire, and a blocked retreat. The gatehouse itself would be a keep in miniature, differing from its predecessors in that it had a tunnel entrance running through it. In this form it might be known as a barbican. It would be three or four stories high, rectangular in form but with round or octagonal towers at the corners. Machicolations were always built above the front entrance, and frequently at the rear. As fire was frequently employed to burn down the gate the barbican had chutes as well as machicolations through which water could be poured. Formidable barbicans are to be seen at Caerphilly, Denbigh, Alnwick, and Harlech.

Although the term 'barbican' was widely used there has been some confusion between what constitutes a gatehouse, and what is implied in a barbican. The difference is similar to that between 'tower' and 'castle'; the latter is more elaborate and sophisticated. A barbican amounted to more than a heavily defended gatehouse. Flanking towers, before and behind the main entrance, menaced every point of the immediate approach, of which Denbigh provides a formidable example (Figure 6).

Caerphilly, although in the middle of an artificial lake, and

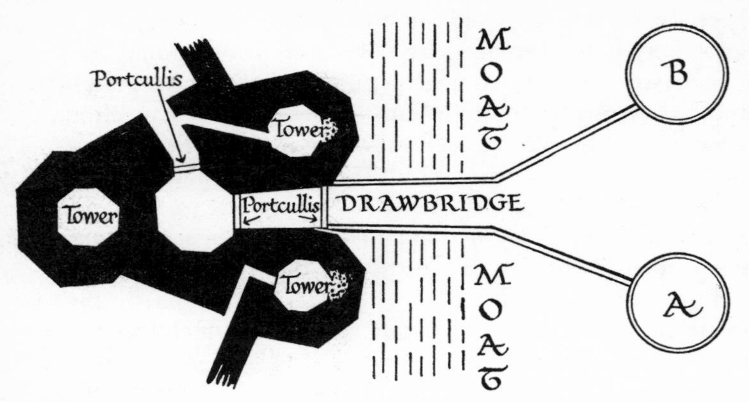

Figure 6. The gatehouse at Denbigh. Note the three portcullises. Built between 1295 and 1322 it was the strongest point in the castle and concentrated the heaviest fire at the doorway.

Towers A and B are imaginary but show how an outer barbican could have been constructed.

otherwise protected by difficult approaches, had one of the most formidable entrances ever designed. It was in effect a castle in miniature. The outlying towers of a barbican could on occasion be some distance from the rest of the entrance but they controlled and even directed the point of entry.

Although castle building reached the peak of sophistication in the fourteenth century, there were many areas that did not require elaborate structures or, if they required them, still had to make do with lesser buildings. In consequence we find there was a return to the simple tower, mainly in border areas. They were known as pele towers, the term being derived from the Latin 'pilum'—a stake. Many of them resembled square church towers which survive to-day, and there is no doubt that in certain areas the church architects had a keen eye for the defensive aspects of these buildings.

During the fifteenth century gunshot developed to the point at which it could batter down the most formidable walls. Some attempts were made to incorporate gunloops in the new or rebuilt castles but adequate defence by gun against gun was soon found to be impossible. Kirby Muxloe, in Leicestershire, provides some interesting examples constructed in 1480. Bodiam and Hurstmonceux adapted oilets to artillery but the problems of gunfire were soon found to be much more than just widening apertures. Ventilation was important but the chief drawback of gunfire was that its vibrations in an enclosed space would in time weaken the surrounding structure. (See Figure 5.)

Before long it was realized that the best defence against artillery was the earthwork, although there was, on occasion, considerable security behind old stone walls.

THE WEAPONS OF SIEGE

Attacking a castle brought into use a wide variety of weapons and equipment, but fire, in various forms, was the most important and most regularly used. In the early stages it was used to burn palisades and wooden towers; later it was employed to crack stone walls, or set roofs alight. The defence found it a most effective weapon against wooden towers (the famous *malvoisins*), that were pushed up to the walls. Its most effective form was the 'Greek fire' discussed later in this chapter.

Arrows were of little use against palisades and, until the crossbow brought longer range, were ineffective against stone structures. In the early days of assault on fortified positions victory could only be gained by heroic hand-to-hand encounters, aided by fire. Heroism alone was not enough against well-defended stone walls so the attack moved underground. The miner was the most feared of all attackers. Sometimes he worked above ground, biting into the walls with pick, spike, or ram, but whenever the nature of the soil permitted he would go deep underground and reappear inside the walls. The skill of the miner was reflected in the number of sites which, otherwise vulnerable, were immune through water to the slow but deadly process of undermining. Considerable subtlety was employed in the underground approach. The entrance would be distant, and well-concealed. Diversionary attacks would be staged to distract the defenders' attention. As nothing could be achieved from the surface the castle-holders would dig out counter-mines, and on several occasions would break into the besiegers' galleries and engage them in hand-to-hand combat. There are numerous accounts of desperate battles underground, and the skill, science, and courage of the attacker was often

Figure 7. A wooden scaling ladder built on the lattice principle.

Figure 8. A scaling ladder made of wood and leather.

matched by similar qualities in the counter-miner. In St Andrews Castle there is a well-preserved mine which, although belonging to 1546, is probably typical of mediaeval operations, although perhaps larger than most, being seven feet wide and six feet high. Also preserved is the counter-mine which had some difficulty in locating its target but which eventually broke into the mine from above and settled the fate of the attackers.

There was much difference of opinion over whether a counter-miner should approach from above or underneath; supporters of the latter theory placed great reliance on their ability to smoke out the miners by lighting fires in galleries below.

Detecting a mine was achieved by a simple but effective method. Jars of water were placed on the ground at intervals; if the water quivered it was due to vibration coming from below. The drawback of this method was that the mine was often not discovered until it was far advanced and already in a position to bring down part of the walls. However, as soon as a mine was detected a fresh palisade was built between that and the rest of the defences.

On the surface, miners were sheltered by a penthouse known as the 'cat', 'sus', or 'vinea'—a timber gallery with a pointed roof, usually of iron. In 1256 one of these was set on fire by incendiaries from above but the miners, with great presence of mind (one can hardly say coolness), pushed it up to the city gate and burnt it, thereby achieving entrance and victory.

Scaling ladders took various forms, some being made of wood, others of leather. The wooden ones were on the lattice principle and could thus be projected on to their objective (Figures 7 and 8).

The ladder was brought under the walls and then pushed into the extended position. The top would be clamped around a merlon while the base would be secured with stakes. It would thus be difficult to dislodge. Anyone exposing himself on the walls in an attempt to unhook the claws would promptly be a target for a dozen well-directed arrows. Climbers would often ascend under the ladder, pulling themselves up by their hands, a feat demanding considerable gymnastic skill.

Towers were used from the earliest times. These were made of wood and pushed forward on rollers. They were given the name 'malvoisins', i.e. 'bad neighbours'. They were also known as berfrois, berefredums, or belfragiums. They were usually three or four stories high, and a more uninviting mobile death-trap can scarcely be imagined. Assault by towers was ultimately effective but the nature of the operation involved a high casualty rate. Being entirely of wood they were particularly vulnerable to fire, which the defenders directed on to them with incendiary arrows. As they were crammed with assault troops the effect of a few well-directed incendiary missiles may be imagined (Figure 9).

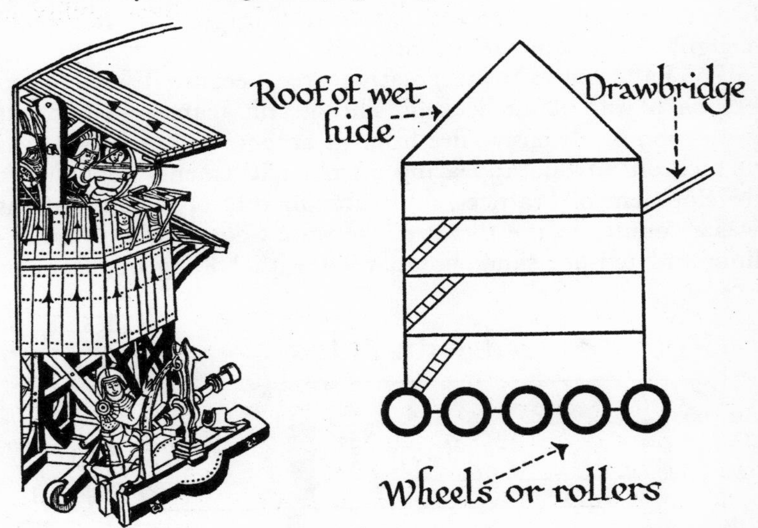

Figure 9. The belfry or assault tower. *Left.* Contemporary drawing of a belfry from British Museum Royal MS. 14, Edward IV, f. 281, *circa* 1480. *Right.* Diagram of a belfry which might comprise several floors to gain the necessary height from which to attack the defenders.

Before a castle could be approached by a tall heavily-laden tower the approaches had to be smoothed, and ditches filled in. On more than one occasion we read of the tower canting sideways at the critical moment, leaving its occupants helpless. Such an event was occasionally produced by careful preparation of the ground: this would be undermined but left strong enough to bear the weight of a company of men; if the company of men were concentrated in a tower of three or four stories the surface would cave in. This method was first used at Rhodes. Care had to be taken to prevent the tower falling forwards as this would aid the attacker. The largest tower recorded was the one Richard I produced before Acre in 1191, whose walls it overlooked. He named it Mategriffin—Checkmate.

Rams and bores were used to batter at gateways or wall-bases. Bores were lighter than rams, easier to handle, but consequently slower to take effect. In the weird humour of the day such a weapon might be called a musculus (mouse) because it gnawed a hole, a cat, because it clawed a way in, or a sow because it bored with a tusk, and when draped with hides had a slight resemblance to a rooting pig.

The battering ram was greatly feared because it would soon breach a wall if unchecked. The assault team relied for its protection on defensive fire from its archers whose duty it was to pick off anyone appearing on the battlements above. The development of brattices, and subsequently of machicolations was a counter to the threat of the ram. Showered with quick-lime and red-hot sand, bombarded with boiling oil, lead, or

Figure 10. The Ram. The troops were protected by a roof covered with hides: the ram could rotate in the slings.

water, the assault team had the least enviable siege task. A ram might be a tree-trunk with a metal tip. It would be slung under a wooden framework, protected by a covering known as a mantlet, and swung back and forth by a picked and specially trained body of men. The defenders would drop heavy stones, and incendiary material onto the mantlet and, on occasion, let down a two-pronged fork to grip the end of the ram and immobilize it. Sacking was also lowered to cushion the impact.

While the mine was starting on its deadly way and the bores were hammering at the base walls the attackers would be softening-up the defence with the equivalent of modern artillery. Their weapons were mangonels, ballistas, and trebuchets.

A mangonel was a simple contrivance of two stout posts with two elastic ropes between them. The ropes were made of plaited human hair, which is highly tensile. When a mangonel was built recently for a television production nylon had to be used as adequate supplies of human hair were not available. A beam is placed between the ropes and twisted. When sufficient torsion had been gained a large stone, perhaps 5 cwt., would be placed on the end of the beam and launched in the general direction of the target. The result, being unpredictable, was often devastating to the target area, but occasionally to the launching site. On several occasions a dead horse was projected

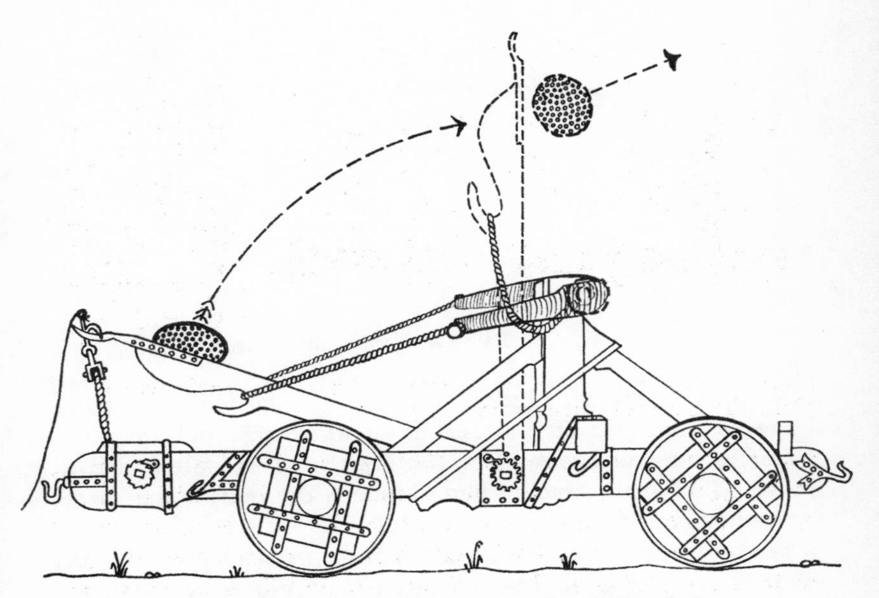

Figure 11. The Catapult or Petrary—after Viollet le Duc.

into the castle, as an early form of germ warfare; on one known occasion an envoy from the castle who had brought unacceptable peace terms was shot back into it with the rejection strapped to him (Figure 11).

A ballista was more accurate than a mangonel, being in effect a giant crossbow. It was also known as a springal. It had

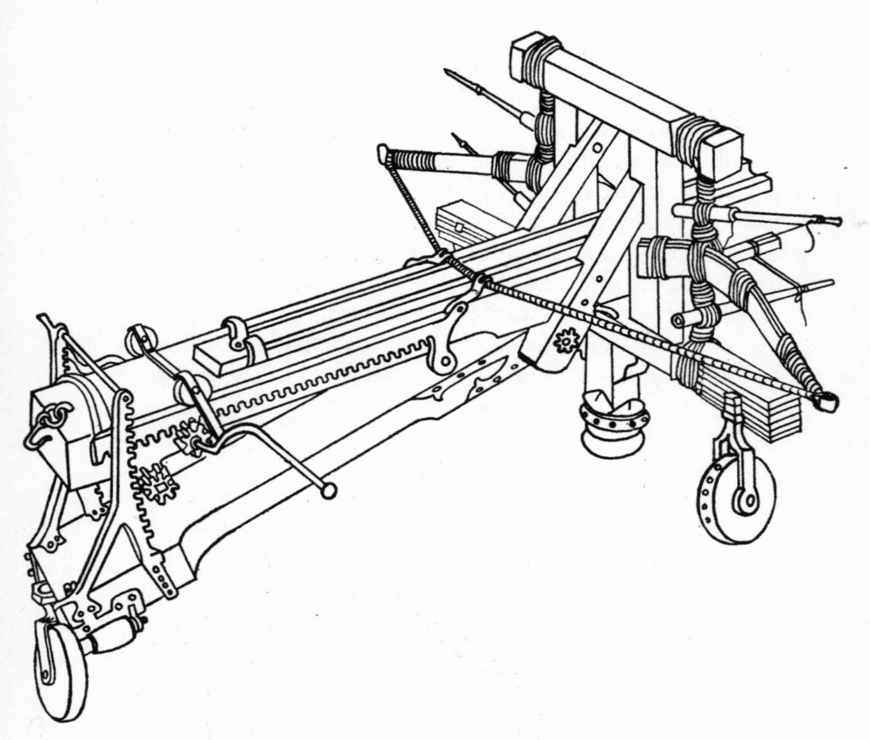

Figure 12. The Ballista—after Viollet le Duc.

little effect on masonry but by its accuracy and force had considerable influence on morale as an anti-personnel weapon. When Rome was besieged by Witges a ballista bolt nailed a Gothic chief to the tree he had climbed; the body hung there throughout the siege.

The trebuchet was a lever on a fulcrum, and proved very effective although it was costly and cumbersome. Heavy weights were suspended from the forward end and these, when the rear portion was released, swung it into the air with its missile. Its drawback was that the trajectory tended to be high and therefore lose force. A petraria was a similar weapon, and perrier was yet another name for it. In 1339 the French launched dead horses by this means into the castle of

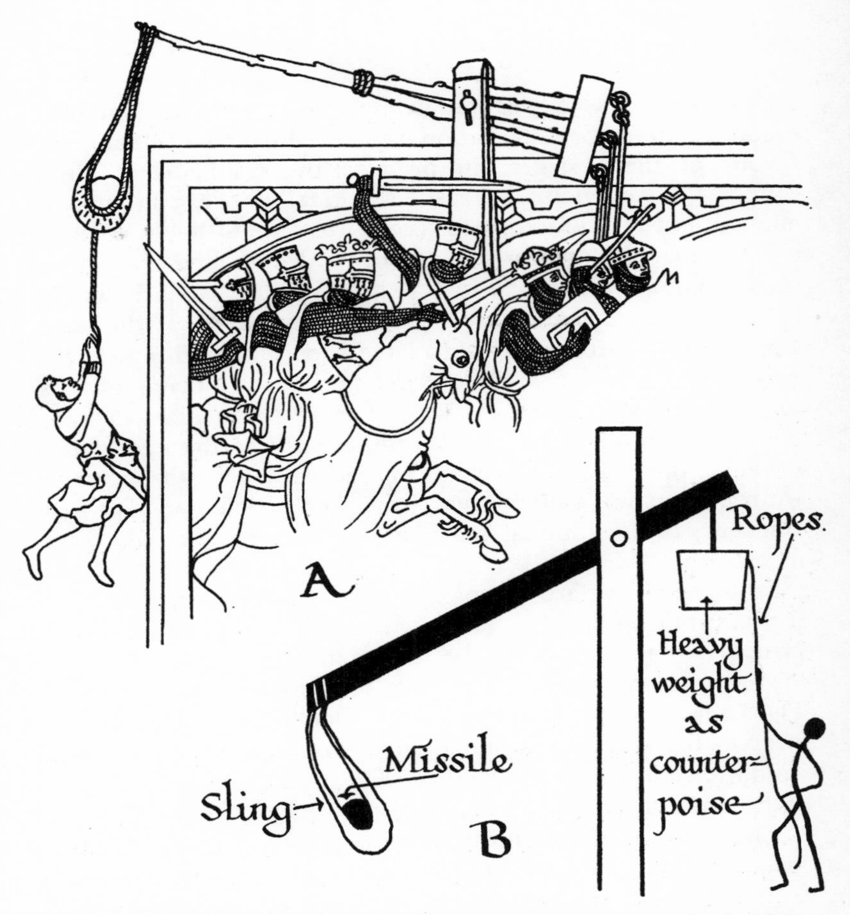

Figure 13. The Trebuchet. *A*. Contemporary illustration of a trebuchet from the *Maciejowski Bible*, French, *circa* 1250, Pierpoint Morgan Library, New York. *B*. Diagram of the trebuchet. This siege engine was used for heavy missiles. Its principal disadvantage was the massive counterpoise which was cumbersome to transport.

Thin. In 1345 a captured messenger was launched back into Auberoche.

The men in charge of siege engines were called 'gynours'. Sometimes, instead of weights they pulled down the arm by ropes; this method was quicker and saved the transport of heavy counterpoises, but was not the most effective.

STARVATION

Starvation was a weapon whose effects were seldom predictable. In theory starvation should always have acted in favour of the besieger; in practice the results were often different. A well-victualled castle would fall if a siege lasted long enough, but on numerous occasions was able to hold out until relief came or a truce was arranged. Impending starvation might activate a garrison to a point of frenzy, when it would break out and defeat the besiegers in open battle. If the siege was in the winter crops would have been gathered and flocks killed; the besieger would therefore have to bring all his supplies into the district. It was, of course, a mediaeval practice to kill off all but a few breeding stock at the end of the summer, there being no means of feeding a herd through the winter. Meat was salted down, fruits were dried, and pigs and chickens, the traditional scavengers, provided the only fresh meat.

In modern warfare campaigns have usually been timed for the autumn when the harvest has been gathered, and the principal objectives could be gained before winter set in. A mediaeval campaign was best timed for the mid or late summer. The harvest would be in the fields, flocks would be grazing, and wooden defences would be dry enough to burn. A castle would have low stocks of food, a well affected by dry weather, and a moat diminished in size.

There were subtle moves in the game of using hunger. Before invading Wales the Normans encouraged rival Welsh factions to fight each other. So fierce was internal rivalry and hatred that little encouragement was necessary, and the results benefited no one but the invader who entered a country where crops had been burnt, animals slaughtered, and defences broken, and where the population was dying of starvation or pestilence. The conditions that had been created in Wales became a part of the English scene during the anarchy of Stephen when plunder, burning, and terror, not only destroyed stocks but also prevented their replenishment; in consequence sieges were of short duration.

Water was even more vital than food. The siege of Exeter, in Stephen's reign, was a spirited occasion when fish and meat were plentiful, the castle was superbly garrisoned, morale was high on both sides, and it appeared that the rebel Baldwin would hold out against his king until the latter called off the encounter. Unfortunately for Baldwin disaster overtook him after three months of vigorous sallies, repulse of night attacks, harassment of the besiegers, and counter-mining; the wells in the castle ran dry.

The absence of water did not at first seem disastrous as wine was used instead, not only for drinking but also for bread-making and putting out fires. The latter practice soon used up the wine supplies, and the besieged, driven to desperation, asked for honourable terms. However, their efforts to conceal their extreme thirst were not convincing and they were sent away from the first conference without achieving anything; subsequently Stephen relented and they were let off lightly.

The only castles that were in a position to endure anything more than a short siege were those with access to a river or the sea. Bristol was impregnable to an attacker lacking shipping and Exeter was in almost as strong a position; Wallingford is described in *Gesta Stephani* as stocked to hold out for years. The latter, of which the outline earthworks and motte remain, was powerfully garrisoned and doubtless maintained its food supplies from the stretch of river it controlled.*

As will be realized, the castle garrison did not sit quietly and wait to be attacked. Spies, often women, were sent to discover the plans of the attacker, and frequent sorties, often at night, destroyed towers and siege weapons. Picketing appears to have been extremely inefficient and little difficulty seems to have been encountered in slipping through the attacker's lines. Matilda's escape from Oxford is the outstanding example.

* Calais (1347) is the classic example of the power of famine in reducing a fortress. All previous attempts to force a way in failed.

Above all, the besieged endeavoured to reverse the roles, so that the besiegers were allowed no respite from harassing attacks. This was excellent policy, for a besieger could break off a siege without loss of face and a peculiar situation might emerge in which both sides might account themselves the victors. However, this type of harassment was only possible after the introduction of posterns and sallyports. It would have been impossible in the early Norman castles which had no doors on the ground floor (a fact very noticeable in the keep at Kenilworth) as the sallying party would have had no means of quick re-entry.

THE BOW, CROSSBOW, & LONGBOW

As mentioned above, the Norman bow, on account of its short range, was not greatly esteemed in siege warfare. Its limitations were due to the fact that it was only drawn back to the chest, and therefore lacked power. However, like many other serviceable but not picturesque weapons, it played a more important part than is generally acknowledged. It was effective enough to decide the issue at the Battle of Hastings, though there is some doubt about the alleged arrow in Harold's eye. In Stephen's reign there are accounts of attackers under a close cover of arrows. The crossbow had clearly made its reputation by 1139 when banned by the Church at the Lateran Council. The fact that it was banned did not entirely prohibit its use but it was resisted by common consent among knights for it removed an advantage they had in war. Bows were, of course, only one method of projecting missiles. Slings were also used with great effect.

Accounts of the Battle of Crecy give the impression that the crossbow was an inefficient weapon, but this view is totally inaccurate. Crecy was lost for a number of reasons, one being that a shower of rain wetted the crossbow strings of the Genoese. But the crossbow remained in use long after Crecy, and was an extremely powerful weapon. The longbow largely replaced it because the longbow was suitable for a wide variety of purposes and had a rapid rate of fire. The longbow differed from the Norman bow in that it was drawn back to the ear whereas the Norman bow had only been drawn back to the chest. It was apparently used by the southern Welsh in the twelfth century.

Although not mentioned at all in the Assize of Arms of 1181 it had become the national weapon of England by 1275.

The English longbow was five feet long, and made of yew. It could discharge an arrow 240 yards at a rate of twelve a minute. Every one of those shots would find a target; which was usually a knight's horse. The trained bowman removed all his arrows from his quiver and put them on the ground by his left foot: a body of archers in action would appear to be bobbing up and down like a troupe of gymnasts.

Although 240 yards was the standard range it is probable that many bows were capable of longer ranges. The present British record is just over 500 yards; the world record which was made with a foot-braced specially-built bow is nearly 940 yards.*

So lethal was the sharp barbed arrow of the longbow that an immediate effect of Crecy was the widespread use of the pavise, or pavas; this was a large portable shield that protected the knight and his valet. The knight continued to carry his personal shield; the effect was to make him a castle in miniature for he also wore his hauberk, surcoat, and breastplate.

As the longbow reached the height of its fame the crossbow also developed. The earlier models had been on the lever principle, but later more powerful models, using a form of windlass came into use. Such weapons had a range of about 350 yards, half as much again as the longbow.

Although a powerful and accurate weapon, the crossbow had the disadvantage that it occupied the entire attention of its user. The longbowman, by contrast, could give most of his attention to watching the target as reloading was quite a simple operation. The crossbow was heavy but it occupied a wide frontage; the longbow needed a very narrow front only and thus permitted close and concentrated formations. But for a longbow you needed a strong man; a boy could manage a crossbow.

* According to the 1966 Guinness *Book of Records* the world record bowshot was made by Sultan Selim III in Turkey in 1798, and was a distance of 972 yards 2¾ inches.

The modern record is 937·13 yards made by D. Lamore of U.S.A. in Pennsylvania in 1959, using a foot-braced 54-inch maple and glass-fibre bow with a 250-lb. pull.

The British record is 507 yards 1 foot 1¾ inches made at Radley, Berkshire, by R. Bamber in 1964.

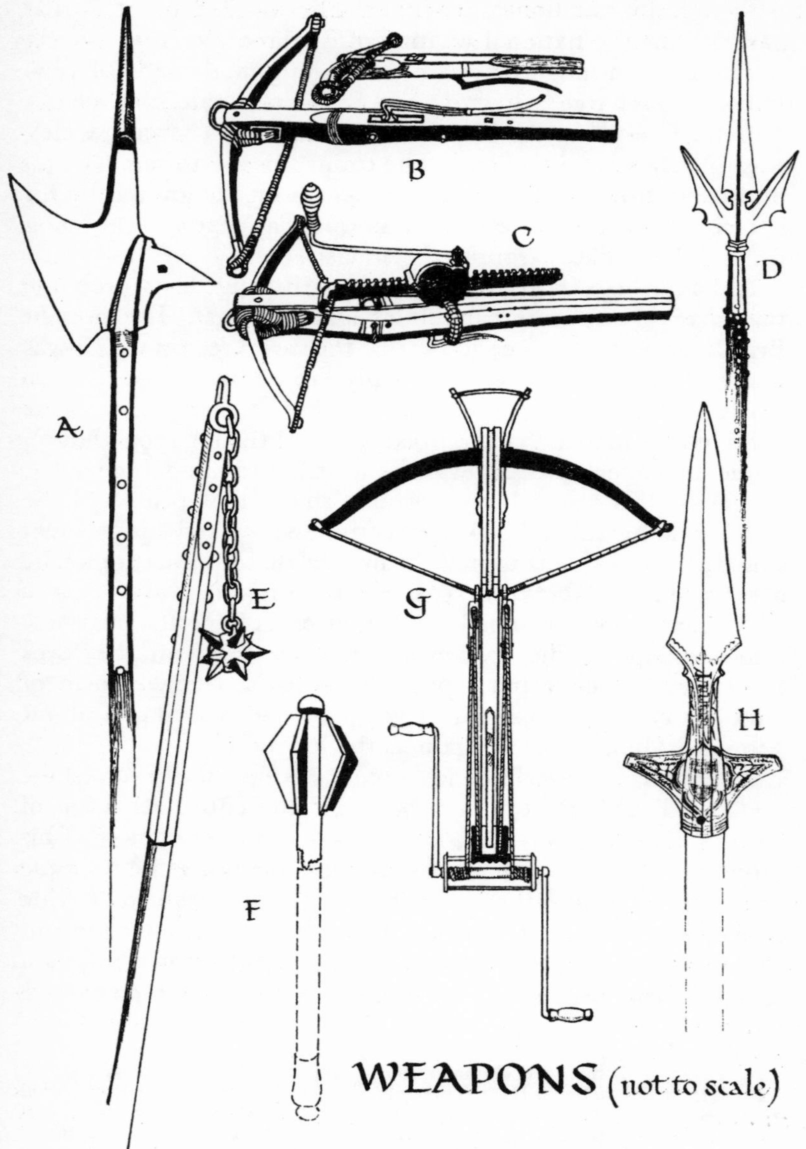

Figure 14. Mediaeval Weapons.

The Rolling Purchase crossbow also worked on a ratchet principle; the strings were wound up from handles on the sides.

An unromantic but extremely useful weapon was the flail. This was a club with pieces of short chain attached to it. Sometimes each chain had a ball on the end. A blow with a flail was virtually impossible to parry but could easily crack open plate armour and stun the wearer in the process.

Accounts of siege warfare lay great stress on the work of military engineers, and too little on the ingenious and courageous people who did the close-quarter fighting. The primary requirement among the latter was agility and speed. They were required to swim moats, scale lofty walls, fight on precarious ledges, and always be alert for an opportunity to outflank the defence. They were lightly armoured and lightly

---

A. Halberd, German, *circa* 1500. No. A.952 in the Wallace Collection, London. This halberd is typical of the kind carried by the German and Swiss Infantry at the end of the fifteenth and during the first half of the sixteenth centuries.

B. Two composite lever crossbows, the upper one from above, the lower one seen from below showing the lever. Second half of the fifteenth century.

C. Crossbow with cranequin (ratchet and lever) in position. *Circa* 1560.

D. Corsèque, early sixteenth century. Similar to No. VII-853 in the Tower of London Armouries.

E. Military flail, Bohemian, fifteenth century. Length of the shaft approximately 6 ft. (The examples, occasionally seen with short shafts, like maces, are fakes: a man wielding such a short-shafted weapon would find that the ball would swing round to hit him first!)

F. Mace. Iron mace head from London, fourteenth to fifteenth century, London Museum No. A, 1778. (The mace is not necessarily a bishop's weapon. Contemporary illustrations show it being used by knights and men-at-arms.)

G. Windlass crossbow. The bow was bent by the windlass and its bowstring secured by the fingers of the nut. From *The Crossbow* by Sir Ralph Payne-Gallwey, London, 1903, Figure 76.

H. Spearhead made for the Emperor Frederick III (1415–93) before he became Emperor in 1439, Kunsthistorisches Museum, Vienna, No. A.32.

armed; the dagger was their universal weapon. Daggers were usually the weapons of poor and despised troops whose only use would be to despatch the fallen or the wounded, but in sieges the dagger men were forward assault troops.

Other weapons used in close-quarter fighting, around gateways and breaches in the walls, were weapons which could be pushed or swung. In the first category came pikes and lances. These might be up to twenty feet long. Variations in the point caused it to be given different names but its function was essentially the same.

Halberds and pole-axes were deadly weapons but required space for manœuvre. Properly swung they could cleave an armoured knight to his chest, or even further. Anyone who has seen a woodman or hedgecutter swinging axe or slashhook will know how nimbly and expertly such deadly weapons could be used.

The halberd which was eight feet long with a hook at the back and hatchet at the front was a particular weapon of the Swiss mercenaries; it was superseded by the morning star, a five-foot club studded with spikes. Both weapons were used by English soldiers.

All or any of the foregoing might be in use at a siege for there was seldom any co-ordination of attack. Mediaeval organization and discipline was chaotic, being based on rank or size of contribution of arms and men. Squabbles broke out frequently and rivalry was sometimes so great that the defeat of an ally was viewed with satisfaction. Troops consisted of knights, esquires, mounted archers, foot archers, billmen, gynours, and pavisers. Crossbowmen were usually hired Genoese. They fired quarrels—arrows with four-sided heads. Fifty was the quota for every crossbowman but their weight made it impossible for more than eighteen to be carried by one man. A troop of knights would be commanded by a banneret. Knighthood was determined by property: if a man had more than £20 a year he was liable for knight service. However, it was an office of distinction and had a high social precedence. Light cavalry were called hobilers and took their name from the fact that they rode hobbies (hobby = a small horse). Billmen swung axes, halberds, partisans, or other similar weapons. Pavisers had be be active for it was a common custom to shoot an arrow attached to a string, aiming at the pavis or mantlet; the pavis was then dragged over while a comrade put in a swift shot to the exposed target. This method was first used at the siege of Roche-au-Moine in the thirteenth century. Pavises were also used to block breaches made by missiles in battlements, to cross marshy ground or moats (Froissart) or for protecting the heads when marching through hostile streets (Ypres 1383). Although the feats of the longbow at Crecy made the use of the pavis universal the accuracy of crossbowmen had brought it into partial use much earlier. At the siege of Brest 1388 the Genoese crossbowmen showed an accuracy worthy of William Tell. Every head that appeared above the battlements was promptly transfixed. We do not hear of the shots that missed, but the quantity of bolts, arrows, or quarrels used was generally enormous.

The William Tell story in which the father shoots the apple from his son's head has, like many stories for children, a slightly sinister setting. It is possible that all members of certain Swiss Corps of archers had to undertake such tests, designed both to test accuracy and to show that personal interests must rank second to martial skill. It has been claimed that archers were pagan and fired at the centre of crosses in practice. Undoubtedly, as the Robin Hood story shows, they had at least a touch of paganism in a code of laws that was developed from a military mystique. Inns called 'The Green Man' exist in former forest areas. There is one at Coleshill (Warwickshire), once a heavily wooded district.

Nowadays we celebrate April 23rd as St George's Day but in earlier times it was Green Man Day, an occasion as pagan as Saturnalia, which preceded Christmas. Pagan religions linger on in forests and remote areas. Fear inclines woodland dwellers to animism, and at the same time they are remote from town influences. It is easy to be briskly sceptical about superstition when one is sitting comfortably in a town house, but a different attitude may prevail when a man is lost in a deep forest or stranded on a remote mountain: at such times misfortune seems personal.

To draw a longbow required a pull of 70 lbs., hence the archer needed strength and constant practice. In the sixteenth century the bow could still outdistance the arquebus by 100 yards, for that was the latter's entire range, but whereas any weak fool could be effective with a gun, provided it did not jam, you had to be a man to use a bow. A bow did not jam but it was not a weapon for a sick or weak man.

Siegecraft produced a variety of weapons that make a modern infantryman seem lightly equipped. Pride of place must go to quicklime, which was thrown down from walls with deadly effect (it was also a popular weapon in sea fights). At the best it would blind, at the worst choke and burn.

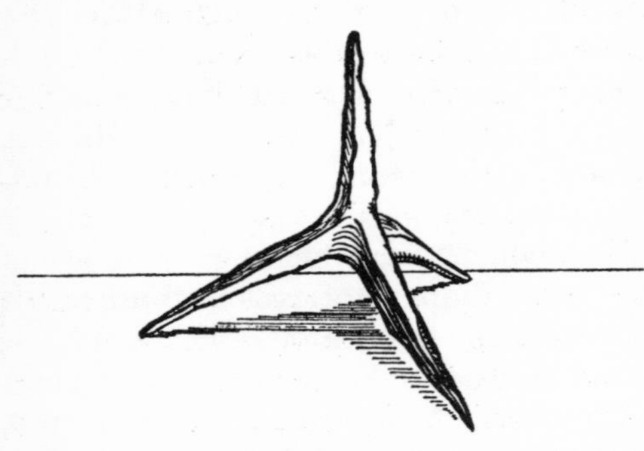

Figure 15. Iron caltrap. Half actual size. Tower of London Armouries, No. XVIII-66.

Red-hot iron bars were another popular tool for they would, with luck, set alight to a pavis or mantlet. Hot stones were used for the same purpose but tended to disintegrate in heating, or on impact. Caltraps were used in thousands. They were iron spikes set so that however they fell one point would always be upwards. They caused appalling confusion when scattered under charging horses.

It was the custom to scatter caltraps on the ground surrounding castles, and particularly on the slopes immediately below walls or keeps.

ARTILLERY

A form of artillery was first introduced in the early fourteenth century but was not as effective as catapults. It discharged balls or darts weighing up to three pounds. The gunpowder was fired through a touchhole in the breach. In 1356 the Black Prince used artillery at the siege of Romorantin; the town was set on fire and capitulated.

By the end of the century vast strides had been made, and cannons could project missiles weighing up to 200 lbs. Smaller cannons were more accurate than their bigger brothers, and helped to win the battles of Tewkesbury and Losecot in 1470; they also sufficed to reduce the northern castles.

The weakness of the early artillery lay mainly in the gunpowder which contained coarse saltpetre and therefore burnt slowly. The only saving grace was that the slow combustion preserved the cannon. When finer saltpetre was used there was almost as much danger behind the cannon as in front of it: if the barrel did not burst open sideways there was a fair chance of the recoil blowing out the back. Lighting the gunpowder through a touchhole soon became suicidal; the new method was to lay a train of quick-burning powder, quick but not too quick to prevent the artilleryman retiring to a safe distance.

The hazards of early artillery were dramatically demonstrated in 1460 when James II was at the siege of Roxburgh Castle. He was particularly interested in a large hooped bombard (a type of gun which consisted of semi-circular plates welded together with hammering and bound with hoops). It was an import from Flanders, had a tremendous reputation, and had been named 'The Lion'. Unfortunately it exploded, killing James instantly and wounding many of his followers.

The weight and cumbersome character of cannons gave great trouble to their users. Once in position a cannon was difficult to move; this was not important when the task was battering a castle wall but a considerable handicap in open warfare when enemy troops could easily manœuvre out of range. Curiously enough, wheels were not used till the late fifteenth century.

Early cannon balls were of stone bound by hoops. Once the principles of casting were understood projectiles became much more sophisticated. Considerable ingenuity was shown in the use of heated shot, hollow balls that burst on impact, and case-shot, but most of these devices were more dangerous to the gun crew than the enemy. When bronze was substituted for iron in fifteenth-century gun barrels it was found to be more enduring; however the copper-tin alloy was not always accurately gauged, and bronze cannon acquired a reputation for bursting without warning.

In the early days of cast-iron cannon balls there was a tendency to make them the same size as the stone missiles they replaced. In consequence, early gunpowder could scarcely move them. The mistake was soon realised, drastic reductions were made in size, and the importance of muzzle velocity appreciated. After this castle walls were doomed.

Between 1449 and 1450 the French successfully attacked 60 citadels in 369 days—an average of a week per siege. In 1453 the Turks captured Constantinople, which had the most sophisticated defences in the world at that time. Although all methods of assault were used there was little doubt that gunfire was the deciding factor. Eleven years later Edward IV battered his way into Bamburgh, formerly thought to be impregnable.

While these larger weapons were being developed an early form of machine-gun—the ribauld—was in frequent use at sieges. A ribauld consisted of several tubes clamped together and swung around so that they would fire in an arc. They were used at Calais and were particularly favoured for attacking breaches or doorways.

Although gunpowder leaps to prominence in the thirteenth century it was probably known much earlier. Roger Bacon first mentions the explosive quality of sulphur, saltpetre, and charcoal in 1249, but the originator of the custom of putting it behind missiles is unknown. Contrary to popular belief the Chinese did not invent gunpowder, although they used powerful incendiary materials from an early date.

The most feared and devastating device was 'Greek fire'. To this day the exact constitution of this terror weapon remains a mystery. It was liquid, could be blown from tubes, would burn on water, and even stone and iron could not resist it. It could be extinguished only by sand, vinegar, or urine (which contains potash). This was the most sophisticated of mediaeval incendiary missiles, but it should not be forgotten that less formidable, but still deadly, materials had been in use for many years. The Ancient Greeks and Armenians used mixtures of pitch, resin, and sulphur, the Romans quicklime and sulphur (which ignited on contact with water), and from then on mixtures of sulphur, quicklime, naptha, essential oils, petroleum, and salt produced the ultimate weapon. There was probably no standard formula but anyone of the latter variants was adequate; whether blown from tubes, thrown in containers or used in mines it was highly effective. An additional use was to heat up sand so that red hot it could be poured on assailants and go through chinks in armour.

There is not space to go into all the ingenious devices which went into castle constructions. Posterns and sallyports, enabling defenders to attack or to escape were constructed to shelter the user. They were more numerous in later castles. Stairways usually turned to the right as they ascended thereby giving the person defending from above the greatest freedom in the use of his sword arm. As the positions might be reversed there are exceptions and at Beaumaris six stairways turn left while four turn right; at Caerphilly seven turn right and two left.

The well was a vital part of a castle and might influence the choice of site. Some of them were extremely deep. Although the water supply was small it would be unpolluted; very little water was wasted in unnecessary washing. Curiously enough if a person does not wash at all (as happened to certain people on campaigns in the Second World War) he does not seem to get particularly dirty, for the skin seems to clean itself after a time, although he may be troubled by vermin. Mediaeval soldiers were probably not smelly or dirty but were undoubtedly itchy.

The knight's mount was not, as is commonly thought, a hulking cart-horse but a much lighter and more active animal. It was not likely to have been sturdier than a heavy hunter, particularly as horses, domestic animals, and people were much smaller at that time. (Sheep for instance were the approximate size of large dogs.) Armour weighed under sixty pounds so a man going into action would be less burdened than a modern infantryman. The greatest drawback to armour was its suffocating effect; even in the cool climate of England we read of knights dying of exhaustion.

Any ideas that an armoured knight wobbled around stiffly or once down could not rise, are completely groundless. In full armour he could turn a somersault, vault over a warhorse, or chimney up a wall (propel himself up between two closely situated buttresses). He could climb the underside of a ladder while wearing a breastplate, or do it one-handed stave-to-stave while dressed in ordinary clothes. He was, in fact, a very fit and nimble soldier, and what made him a difficult adversary was the fact that he not merely loved fighting but could scarcely endure existence when he was not doing so.

❋ 3 ❋

The Castle as an Instrument of Conquest

William I (1066–1087) : William Rufus (1087–1100)

As dusk fell on October 14th, 1066, and the long, bitter day had finally given victory to the invader, William rode to the crest of Senlac Hill and ordered a space to be cleared among the bodies. In the middle he set up the Pope's standard, knelt in prayer, and took a solemn oath to build an abbey; the High Altar would be on the exact place where Harold's standard had fallen. At the conclusion of this impressive piece of ceremonial he ordered a feast and caroused with his followers among the heaps of Saxon dead.

This macabre little drama gave an indication of the Conqueror's mind, and a hint of the future that lay ahead for the English. William was an early exponent of what is nowadays called 'psychological warfare'. On this occasion he was reminding his followers of the rightness of his cause, and the reasons why they should remain loyal to him. He had no illusions about the mentality of his motley army, which consisted of ambitious war-lords and foreign mercenaries. Discipline had been maintained with difficulty when they had waited in France for over a month. Adverse winds had blown constantly and there were mutterings that God was against the enterprise. Accordingly, William, who was a great believer in the effect of holy relics on other people, exhumed the bones of St Valery and had them carried in procession through the camp. At the same time he increased the wine ration. Soon after, the wind changed.

William was harsh, cruel, stubborn, vindictive, and probably cynical, but he was efficient, tenacious, imaginative, and occasionally gracious. When Ivo of Ponthieu struck Harold's fallen body William expelled him from the army for a cowardly deed; and when Hereward resisted gallantly in the Fens the English hero was pardoned, but these were isolated acts. For William was a man of his times. He was the bastard son of Duke Robert of Normandy by the daughter of a tanner of Falaise. (William was not ashamed of his origin and at times began declarations with the words 'I, William the Bastard'. His enemies mocked him for it and during one siege hung out hides on the castle walls and jeered 'Bring out the hides for the tanner'. His mother was married off to a suitable French count.) The way to advance was through courage and skill at arms, both of which qualities he showed from an early age, but his mixed background seems to have given him something else. He was aware that conquest must not only be done but must be seen to be done, that the bravest man may not venture if he thinks fate is against him, and that loyalty in the Middle Ages usually had its roots in self-interest. Like many great generals he was an expert at the occasional flamboyant assumption of being on a telepathic line to God.

Although military strategy appears to vary over the centuries, the principles remain largely the same. Clausewitz defined strategy as 'the theory of the use of combats for the object of war' and tactics 'as the use of military forces in combat', but this description is too narrow. Strategy implies the overall direction of a campaign, but usually has to include politics, economics, and psychology. The organization of war—and this applies to the Norman conquest of England—takes the following pattern. First the policy must be decided, its feasibility considered, resources estimated, and timing reckoned. Here, as at all stages, psychology must be included, for however brilliant the planning it will be useless if there is no general wish to fight. The contrast between the French defence of Verdun in 1916 and their failure in 1940 may serve as an illustration.

Policy must be followed by recruitment and training of men, supply of arms and war material, collection and assessment of intelligence, and establishment of bases. Following these there is deployment and concentration, offensive, manœuvre, consolidation, and control of former enemy territory. There may well be withdrawals and defensive phases. Accurate intelligence, speed, and mobility, will be essential.

This is, of course, no more than a skeletal account of the requirements of the simplest campaigns, and omits many matters which would be vital in modern warfare. But efficient communications and medical services were not omitted because they were unnecessary; under the conditions of the time they were impossible.

Successful generals usually have some favourite form of organization which often ensures victory. Caesar had the cohort, Gustavus Adolphus had concentration of arms, Napoleon the autonomous division and Rommel the Panzer unit; William I had the motte and bailey castle.

William demonstrated his faith in castle strategy by bringing one over in pre-fabricated sections, and erecting it on the shore at Hastings immediately after he landed. According to William of Jumieges, writing in 1070, he erected castles at both Hastings and Pevensey but it is probable that in the latter he merely made use of the old Roman fortifications. The Bayeux tapestry shows the motte being raised; in the background two workmen are seen settling a private difference with shovels. When completed and garrisoned the castle was commanded by Humphrey de Tilleul. From such a promising start it might have been thought that the name of de Tilleul would have become famous in his new country but it was soon eclipsed, and he had to forfeit his English estates. The nature of the unusual crisis that caused his disappearance in the year 1068 is delicately described by Ordericus: 'Some of the Norman women were so inflamed by passion that they sent numerous messages to their husbands, adding that if return were not immediate they should choose others. The lawfully created barons and leading soldiers were in great perplexity for they were sensible that if they took their departure while their sovereign with their brothers, friends and comrades, were surrounded by the perils of war they would publicly be branded as base traitors and cowardly deserters. On the other hand, what were these honourable soldiers to do when their licentious wives threatened to stain the marriage bed with adultery, and stamp the mark of infamy on their offspring.' The gallant Tilleul was one of those to return; heroic in battle but henpecked at home.

Having staged a suitable drama at Senlac, William set to work to make the maximum use of his recent victory. He was well aware that Hastings was more probably the first battle in the campaign than the last, and that the victory so narrowly won could easily be lost again. He proceeded swiftly but cautiously to extend his grip.

The first people to feel the onset of Norman terrorism were the unfortunate inhabitants of Romney. These had, not surprisingly, finished off the survivors of the Norman fleet which had been wrecked there in the previous months. After a swift and thorough massacre William left Romney in flames and set off to Dover.

Dover was one of the few places to possess a well-constructed castle. Unfortunately the hearts of its defenders were not as stout as the walls and through a combination of feebleness and treachery the resistance was negligible. Once in possession the Normans soon forgot their promise to respect property and exercise clemency. Soon half the buildings in the town were occupied by Normans, the rest in flames.

Ordericus gives the following account: 'The Duke then continued his march to Dover, where there was a large body of people collected because they thought the position impregnable, the castle standing on the summit of a steep rock overhanging the sea. The garrison, however, struck with panic at the Duke's approach, were preparing to surrender, when some Norman squires, greedy for sport, set the place on fire, and the devouring flames spreading around, many parts were ruined and burnt. The Duke, compassionating those who were willing to render him their submission, ordered them to be paid the cost of rebuilding their houses, and their other losses. The castle taken, eight days were spent in strengthening the fortifications. While he lay there a great number of soldiers who devoured flesh-meat half-raw and drank too much water, died of dysentery, and many more felt the effects to the end of their days.' Without waiting a moment William set his engineers strengthening the castle and prepared for the complicated task of capturing London. Before he left, however, he encountered a hazard, which, had it come earlier, might have made a considerable difference to his attack on Dover. Some form of gastric upset, usually thought to be dysentery, but more probably cholera, raged through his army. The number of dead is not known, but seems to have been considerable, and the weakening effect on those who survived took some time to shake off. Epidemics such as this played an unwanted but important part in mediaeval campaigning. It is generally assumed that only one disease would be at work at a time, but anyone who has had any military experience in an area where water supplies are contaminated and insects are numerous, will know very well that two or three dangerous scourges may be flourishing simultaneously.

Before long William himself was ill, although he was not immobilized for more than a few days. But as soon as he had recovered sufficient strength he lost no time in setting out on his most difficult and important battle—the siege of London.

London was a formidable task. From the south it was protected by the river and strong town walls, to the east there were dangerous marshes, and in the north areas of thick forest. Here he could lose all he had gained at Senlac. Like all good generals he wished to fight battles only after victory had been won. Furthermore he knew that Harold's forces had been exhausted by their long marches and the desperate battle of Stamford Bridge. If he now tried to carry London he would be putting himself in much the same position as the unfortunate Harold. Sun Tzu, writing in 500 B.C., first defined a principle of war which has subsequently become famous: 'Making no mistakes is what establishes the certainty of victory for it means conquering an enemy that is already defeated.' The drawback to set battles is the possibility of losing. Frederick the Great regarded battles as a last resort when he had failed to outmanoeuvre his opponent. Just before the Battle of Hohenfriedberg he wrote: 'There is no way out, as far as I can see but a battle. In a few hours this emetic will have determined the fate of the patient.' When a general can manoeuvre, besiege, and harry he has a good chance of destroying his opponents' desire to fight. This is precisely what William did in his campaign against London.

His first move was to make a frontal attack from the south, and for this he chose Battersea as being the nearest point to the city's most important buildings. The weakness of his position was soon as obvious to William as it was to the Londoners. The town was full of defiant men whose numbers are said to have included Danes, and troops who had been with Harold at Stamford Bridge but who had not reached Senlac. That this force would not be cowed into submission by the few siege engines he was building, was soon noted by William. If anything, his own position was more vulnerable than theirs, for campaigning and disease had reduced his numbers, while pillaging and devastation by his own men had diminished their sources of supply. Before the situation became worse William set off south west in the direction of Guildford to take stock and organize a more subtle assault on the capital.

According to Forester, the commander of the London garrison was one Asgar who had an infirmity of the loins and had to be carried everywhere in a litter. In spite of this he was more than a match for William, saw completely through the latter's overtures, and was clearly as brilliant a diplomat as he was soldier.

As the Normans advanced they secured their route with motte and bailey castles. Although the motte by itself was a formidable defence it was made even more so by a palisade of stakes built round the outside of the ditch. These defences frequently extended well beyond the line of the ditch and before long a second or third ditch would appear outside the palisade. After the first dangers had passed most of these inner ditches were filled in, while the outer one was deepened and widened.

That the motte was regarded as an instrument of the flexible offensive was clearly shown by the positioning of William's castles on his march to encircle and thus besiege London. The modern equivalent is to capture ground and dig in. If there is a counter-attack it is contained and then the offensive is resumed.

Having rested his army at Guildford and taken Winchester into his orbit William decided to press on with all speed. Winter was now approaching, and the Norman army had to move before the winter rains made the route too difficult.

He advanced in two columns, one swinging as far west as Ludgershall, where the motte is still well preserved, the other driving up past Newbury to Wallingford where it established a strong camp and crossed the river. The policy of murder, devastation, and terror was meticulously followed. Although this did not influence the English leaders it undoubtedly affected the morale of those who lay in its path, including, as reports went ahead, the citizens of London. With William at

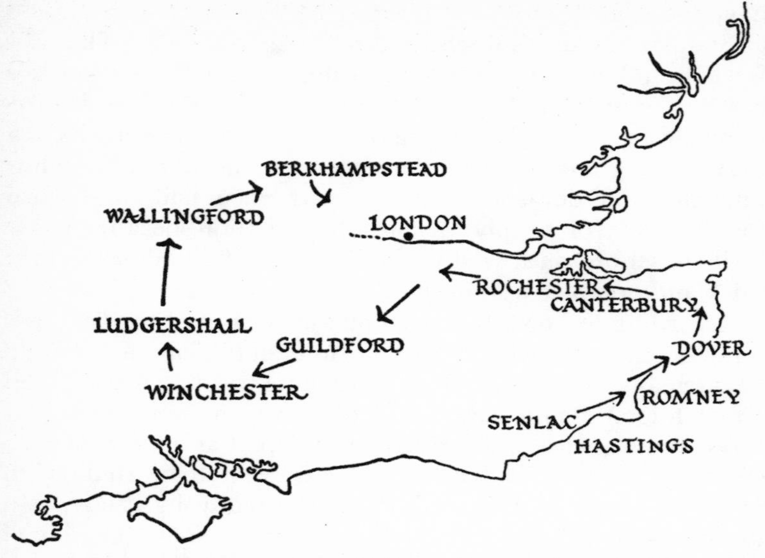

Figure 16. The strategy of William I's campaign of 1066.

Wallingford they could not be sure that the northern English would not abandon them to their fate and give first consideration to the defence of their own territories; in this their fears were soon shown to be justified.

With a secure base at Wallingford William was now able to prevent supplies reaching London from the West. The effects of this devastation in the south were also being felt, although of course it was unlikely that the city would ever be starved into surrender. Terrorism served a dual purpose; it satisfied the ambitions of his motley army, and at the same time it made life more difficult for those of the English who still had a will to resist. Three centuries later it was used in Normandy as the policy of 'havoc'.

But William was not having it all his own way. Marshy land, thick woodland, and local defences combined to make progress hard and uncertain. But finally he reached the position he needed. By December he was twenty miles from the north of the city, now virtually defenceless, and he established headquarters at Little Berkhamsted and awaited results.

He had not long to wait. In the middle of the month a

deputation arrived from London, asking for terms, but prepared for unconditional surrender. They offered him the English crown, and after a suitable though farcical show of reluctance he assented graciously. The first siege was over. The second stage of the campaign was now complete but no one knew better than William that hard fighting still lay ahead.

As a first step in the next stage William decided to stay where he was until crowned, and in the meantime build a castle in his capital. As far as is known he sent men forward to construct a motte on the site of the present Wakefield Tower in the Tower of London. The Motte did not take long to construct, and while it was proceeding a stronger and more elaborate fortress was being prepared; this was the present White Tower.

Mottes, while they had the advantage of rapid construction, had the disadvantage that they could not support stone structures, as these would be too heavy for artificial mounds. Natural mounds, however, are not always sited in the best defensive positions, although the Round Tower mound at Windsor, long thought to be artificial, is an exception.

It must also be mentioned at this point that local conditions sometimes prohibited the building of mottes at all. In the west timber was scarce and stone plentiful so from the outset the Norman castle was of stone.

For reasons that do not concern us here William returned to Normandy three months after his coronation, leaving his half-brother as Regent in his absence. This period, which lasted eight months, was marked by numerous insurrections, which is hardly surprising in view of the way the Normans behaved. 'The chiefs of inferior rank who had the custody of the castles, treated the natives, both gentle and simple, with the utmost scorn and levied on them most unjust exactions' (Ordericus).

Some of the young men went abroad and enlisted as mercenaries under Alexius, Emperor of Constantinople, where they were able to fight Normans on more equal terms. The Varangi, as they were known, were a highly paid, privileged elite who fought with battle-axes. The force was made up of Danes, Norwegians, and English, and latterly became almost entirely English. Varangi meant warrior-band in old Norse; the family

Warenne, former Earls of Surrey, came from Varenne in Seine-Inférieure, and other place names in Normandy derive from the same origins.

Misery makes strange bedfellows, and at this time caused the English to turn for help to Eustace, Count of Boulogne, who was known to have quarrelled with William. Eustace mounted a surprise attack on Dover Castle, coming across the channel by night, but his move was so swift that it defeated its own purpose; and local support, though considerable, was not at its maximum. The siege was desperate but short. After some hours of bitter fighting, Eustace, apprehensive of a sally, gave the order to retreat to the ships. The retreat was turned into a rout by the garrison who opened the gates and attacked the rearguard with disciplined venom. 'The fugitives, panic-struck by a report that the Bishop of Bayeux had unexpectedly arrived with a strong force, threw themselves in their alarm among the crevices of the perpendicular cliffs, and so perished with more disgrace than if they had fallen by the sword. Many were the forms of death to which their defeat exposed them, many, throwing away their arms, were killed by falling on the sharp rocks, others, slipping down, destroyed themselves and their comrades by their own weapons; and many more, mortally wounded or bruised by their fall, rolled yet breathing into the sea; many more, escaping breathless with haste to the ships, were so eager to reach a place of safety that they crowded the vessels till they upset them, and were drowned on the spot.'

On his return from Normandy William set about the second phase of his conquest with the craftiness that had worked so well in the past. He was all things to all men, calming London and the south with promises he had no intention of keeping, and at the same time moving in fresh Norman governors in case the former ones might be mellowing in their attitude towards the English.

With the rear areas thus stabilized he set off briskly westwards with the usual policy of fire and slaughter. His first objective was Exeter, which had made considerable preparations for the inevitable attack. Among these were strengthening of fortifications, detaining of foreign seamen, and the organization of an anti-Norman league of towns.

As William approached, the city-elders began to have some

doubts about the feasibility of resistance, and offered tribute and obedience. They visited his camp four miles from the city, and left hostages. However, by the time they returned the remainder of the inhabitants of Exeter had changed their minds: the policy was now no surrender.

William was surprised and angry, but did not lose his temper. With five hundred horse he advanced to reconnoitre, and finding gates shut and walls manned, moved up his whole army.

He then staged the usual drama by tearing out the eyes of a hostage immediately in front of the city gate, but the spectacle failed to move the Exonians. So battle began.

Accounts of the siege all emphasise the toughness of the fighting. William attacked from all sides, and made desperate attempts to undermine the walls. In the course of eighteen days determined assault he appears to have lost half his army. At the end of that period the citizens decided that surrender might now be opportune, and were pleasantly surprised to find that William, in one of his rare moments of chivalry, respected their courageous resistance, and was very mild in his peace terms. There was no massacre, and no plunder; instead the citizens were granted their lives, property, and privileges. Subsequently their lives were made insupportable, their property taxed away, and their privileges ignored, but their fate, nevertheless, was much better than most people managed to obtain from the Conqueror. William continued to Cornwall and then left the west. His confidence was not misplaced. When, later, Harold's two illegitimate sons came over from Ireland and hoped to rouse the west, their efforts were unsuccessful. With considerable skill William had extracted the maximum value from the Battle of Exeter.

Meanwhile, in the remainder of the country, the effects of Norman tyranny were beginning to show. The English, who had hoped matters would get better rather than worse, were soon driven to desperation. Property was seized from its owners, whether friendly or hostile, and given to William's supporters and favourites. As usually happens, the rearguard of the army was far more ferocious than the men who had done the fighting. There was no hint of chivalry; the English were regarded and treated as serfs, whatever their former rank or standing. Many took to the woods and lived as robbers; others fled to the north where there was still hope of further resistance. The centre of this was the fortified city of York.

Swiftly and with masterly skill the Normans drove north. Oxford, Warwick, and Leicester fell with little resistance, Derby, Nottingham, and Lincoln soon followed. And then, as he encountered the solid core of resistance in the shape of the armies of Edwin and Morcar, the old techniques were again brought out. William affirmed he did not wish to conquer by force, but by fairness and consent. Once more the formula worked; he was believed, and his passage to York was unopposed. The keys were surrendered without a blow being struck; and the Normans, jubilant but wary, fortified the city and built a castle.

William remained suspicious, and not without cause. He was well aware of the dangers that surrounded the network of castles that he had built to maintain his line of communication. What he did not bargain for was that at this point he would lose not only his English supporters but also some of his Norman nobles; these moved up to Scotland where Malcolm Canmore was a generous and undemanding king. But William was practical in his fury. To replace them he brought in a fresh host of supporters, some of whom came from such distant areas as the Rhineland and Italy.

But affairs in York were not going to plan. The castle, far from being a control centre, was virtually under siege. By the time William heard the news in the south the siege was more than virtual; it was taking place physically. With his main pivot in the north threatened William had no doubts about the action required. He travelled in person with a strong force, relieved the city, and stayed long enough to see the foundations of a second castle being built. It had been an awkward moment but speed and severity had overcome the crisis; William returned south to a little well-earned relaxation.

But the north is a large area, and its inhabitants were a mixture of many warlike peoples. The first hint of this came when Robert de Comines set out from York to capture Durham. This was achieved but the ensuing counter-attack by night was so vigorous that only two of his five hundred Normans survived. And this was only the beginning.

During the next two years ominous reports filtered down

from Scandinavia. William was already aware that the King of Denmark considered himself the rightful heir to the English throne and might well support his claim by invasion. But time had passed and the threat did not materialize; perhaps it never would. Complacency was shattered when 240 Danish ships sailed south, some harrying the Channel ports but most landing in the Humber. Meanwhile Malcolm Canmore was reported to be on the march, and the north was aflame.

But, except for the Normans at York, it was nothing but a superbly staged anticlimax. The King of Scotland never appeared and the campaign that might have swept William and his followers back into Normandy never materialized. The Danes eventually found their way back to Denmark; to them it had been no more than a foray, soon over, soon forgotten.

But for the Normans who had been caught in York it was a different story. Outnumbered, they fought with the desperation of the doomed. The walls remained intact for the first seven days of the siege but on the eighth the ditches were mostly filled in, and the end was near at hand. Setting alight the city the Normans burst out, perhaps hoping to fight their way to safety through the general confusion. But the besiegers were too numerous, too thorough, too full of hatred. Of the three thousand Norman troops in York only a few remained to tell the tale, and those only because of the ransom they would bring. The victory at Senlac had cost less Norman blood, and this was a defeat which, if followed up, might well have changed the course of English history. But victory was allowed to ebb away.

William's reaction is aptly described in the well-known phrase 'the devastation of the north'; but the cruelty of the aftermath should not obscure the skill of the campaign which made it possible. He received the news while hunting in the Forest of Dean. Local insurrections in the Midlands, stirred up by the news from the north, hampered his progress. In the west Exeter was temporarily besieged by Cornishmen; these were put to rout by a sudden sally; as at Dover this was a good example of attack being the best form of defence. Shrewsbury was less fortunate. Besieged by men from Chester and Wales it was burnt to the ground. By the time William arrived at York it was winter and a severe one, but nothing was going to stop the cold fury of his vengeance. He ordered the complete destruction of every life, and of everything and anything that could support human beings. Every house and implement was burned; every part of the countryside laid waste. Most of the victims were innocent but this made no difference. William was making an example, and cruelty was absolute. Even Ordericus, usually an admirer of William and his ways, is revolted by the thoroughness of the destruction. 'His camps were scattered over a surface of a hundred miles, numbers of insurgents fell beneath his vengeful sword, he levelled their places of shelter to the ground, wasted their lands, and burnt their dwellings with all they contained. Never did William commit so much cruelty; to his lasting disgrace he yielded to his worst impulse, and set no bonds to his fury, condemning the innocent and guilty to a common fate.' The result was a nine-year famine that affected a wider area than Yorkshire.

Savage though William and his followers were, certain extenuating factors should be borne in mind. The Normans were invaders in a country and at a time when kindness, if it had occurred at all, would have been thought of as weakness. The policy of murder, burning, and terror, was designed to create a state of submission—and did. The Black Prince used the same policy in France several reigns later. To the Normans, who belonged to an organized society the Northumbrians—and the Welsh who were massacred on an equal scale later—were scarcely distinguishable from wild animals. Destruction was more personal than it became later but the bombs and rockets of the Second World War did not discriminate between age and sex either.

With the north 'settled' William determined to teach a lesson to the Welsh border. At this point he encountered some opposition from his French troops who complained 'that they were ground down with a service more intolerable than that of guarding castles, and made vehement claims on the king for their discharge'. They had recently marched from the Tees to Hexham in the dead of winter through severe frost 'but the troops were encouraged by the cheerfulness with which he assaulted all obstacles'. However, William ignored their protests and 'with unwearied vigour made his way through roads never before travelled by horses, across lofty mountains and deep valleys, rivers and rapid streams, and dangerous quagmires in the hollows of the hills. At times they were reduced to feed on the flesh of horses which perished in the bogs. The King often led the way on foot with great agility, and lent a ready hand to assist others in their difficulties.'

Eventually 'to mark his displeasure with those who had threatened desertion he detained them forty days longer than their comrades, a slight penalty for men who deserved a much severer punishment' (Ordericus).

Subsequently Wales became thickly studded with castles. After the initial share-out, Wales and Ireland were the only areas available for land-hungry Normans who had missed the first prizes.

The final stage of William's campaign of conquest was perhaps the most complicated of all. Hereward the Wake bitterly resented the fact that his land had been given to a Norman, and at the first opportunity recaptured it. Well aware that this action would not go unmarked by William, he set about assembling a force large enough to resist attack when it came. He was overwhelmingly successful and soon mustered an army that would have been formidable in any conditions. In the Isle of Ely, where he had a fortified camp, it was apparently invincible. The surrounding country was a mixture of swamp, river, and bog. To the Normans who could not use their cavalry and were unfamiliar with the countryside, it was a nightmare in which they suffered continuous losses from an apparently superhuman enemy. Hereward's supporters knew every available path, and the lighter arms, which had been a disadvantage up till now, at last came into their own. Furthermore, in defending the Isle of Ely, they were fighting for a religious shrine, with all the advantage in morale that such a cause bestows.

William, who had mastered the west and north in the two years 1068 and 1069 found this campaign a more long-drawn out affair. Realizing ultimately that a frontal attack could not succeed because it could not be properly mounted, he resorted to blockade. He collected his navy in the Wash and gave strict orders that the outlet to every creek and river should be watched; and gave the army a similar task on the landward approaches to the Fen country.

As all attempts to find a satisfactory pathway through marsh and river had proved useless, William decided to build his own. The enterprise proved to be exceptionally difficult. The marshes swallowed up material to an extent the Normans had not contemplated and the rivers were difficult to bridge in the face of violent assaults by the besieged. The aim was to build a wooden causeway two miles long but when the marsh was wet it sank, and when the reeds were dry the English set them on fire and burnt it. Small wonder that the Normans began to believe that either God or the devil was against them.

The network of narrow paths enabled the English to attack without warning by day or night, on the flanks, before, or in the rear. Marshy sites are difficult enough on their own as we know from the tributes paid to Radcot, Cricklade, Leeds, Bourton, and Boarstall, but when used for offensive defence as Ely was they are a nightmare to the invader.

But the blockade proved more effective than the assaults. The monks of Ely, whose hearts were nearer to their stomachs than their heads, decided they could end their hardship by treating with the Normans. Unfortunately they knew the secret pathways, and to their shame, and ultimate discomfiture, guided the Normans in.

Once in Ely the Normans had little difficulty in capturing the other key points of Hereward's stronghold. The tale of brutality is characteristic. After the initial slaughter the remainder of their captors were merely immobilised. Some were blinded, but most were maimed for life by having a hand or foot cut off.

But Hereward had slipped through the net. With a few faithful adherents he continued the defence of his own homeland in Lincolnshire, and for several years harassed the Normans with brilliant guerilla warfare. When it was clear to both sides that prolonging the struggle could benefit no one, he accepted the terms William offered. To those who commanded his respect William could be both generous and honest. (The word 'generous', incidentally, derives from 'genus' and denotes a man of breeding and quality.) Hereward was allowed to keep his family estates, and subsequently campaigned in France with William. However, the inaccessibility of Ely was to prove a thorn in the flesh of others later.

In the seven years since William had arrived at Hastings the Normans had been campaigning, in large or small degree, without cease. By 1073 many lessons had been learnt and much blood had been shed but William's position was not much stronger than at the beginning. England seethed under the tyranny, Scotland and Wales were watchful and ready, while across the Channel came reports that trouble was brewing for Normandy. As soon as he set off to deal with the latter an unexpected blow occurred in the revolt of the Norman barons in England. Fortunately for William it was a half-hearted rebellion, and by the time he returned, his loyal supporters had suppressed it. But it was an ominous reminder of the insecurity of his position.

His determination not to relinquish the Dukedom of Normandy soon brought him into bitter conflict with his eldest son Robert. Robert gained the support of King Philip of France and established himself in the castle of Gerberoy, from which he plundered the adjoining Norman territories. William collected a large army and decided to teach his son a lesson. The ensuing siege is of no great importance except for one incident. Robert was not content to be shut up in the castle but led a series of desperate attacks on the besiegers. In one of these he encountered another armoured knight, whom he knocked from his horse. According to some he recognized the voice shouting for help, according to others, as he removed the helmet to give the death blow he saw his father's face. There are other versions also. Although this meeting did not reconcile father and son it indicated the need for some means of identifying people clad from head to foot in armour. Distinguishing marks were soon painted on shields, and helmets were adorned with symbols denoting rank. Subsequently the Angevins wore a sprig of broom on their helmets, and this 'planta genista' gave them the surname of Plantagenet. Heraldry, which later became an elaborate and often fanciful art, had as sternly practical a use as flashes and badges have in a modern army.

Advancing years may mellow a man or make him sourer and more harsh. For William it was the latter. His decrees became harsher, and as the chronicler put it 'the whole country groaned under his yoke'. The accident that led to his death, when he was on a mission of vengeance, was as bizarre and violent as his life. No one mourned him, not even his two sons, who only attended his death-bed to hear his will, and did not wait for his last breath, or take any care for a befitting burial. In the event the great Conqueror was lucky to be buried at all.

Nine hundred years later it is easy to criticize William for his callousness and indifference to human suffering. In the 1960s in a country which is indifferent to six thousand road deaths a year, not to mention a further twenty thousand injured, these condemnations have a slightly false note. At least William did not claim to be humanitarian. As a military commander he ranks with the great captains. He held together a motley army by example and superb leadership. Towards the close of his reign his Norman governors gave him more trouble than the English they were supposed to be controlling; a development that did little to mellow his character.

His English subjects aroused in him the feelings we entertain towards our dairy herds to-day. Provided they produce in increasing quantity we regard them with benevolent indifference, pausing merely to wonder how to squeeze an extra drop for the same quantity of feed; but let them fall off in their output or round on their masters and they are off to the slaughterhouse in double-quick time.

The Anglo-Saxon Chronicle describes him as 'a very wise man, and very powerful. He was gentle to the good men who loved God, and stern beyond all measure to those people who resisted his will. Amongst other good things the good security he made in this country is not to be forgotten—so that any honest man could travel over his kingdom with his bosom full of gold; and no one dared strike another, however much wrong he had done him. And if any man had intercourse with a woman against her will, he was forthwith castrated.'

However the chronicler does not omit to point out that 'certainly in his time people had much oppression and very many injuries'. This was undoubtedly their own fault for as he says of 1087, 'it became a very severe and pestilential year in this country. Such a disease came on people that very nearly every other person was ill of the worst of diseases—high fever, and that so severely that many people died of the disease. Afterwards, because of the great storms there came a great famine. But such things happen because of the people's sins, in that they will not love God and righteousness.'

In the twenty-one years of William's reign England became studded with castles. He was reported to have said that if the country had had an adequate system of castle defence the Norman Conquest would have been impossible. This is probably true.

The early mounds were soon replaced by more permanent structures, but the general pattern remained the same. Height was used to give dominance, thickness of wall was employed for strength. The stone towers were square and often had walls up to twenty feet thick. The Normans had no understanding of the niceties of architecture, and their buildings were unnecessarily solid.

At this period, and for some time later, defence was conducted from the battlements. There was no value, only weakness, in arrow loops lower down. The principal weapon used against the wooden palisades and towers was fire, but when stone replaced timber the possibilities of bringing down the structure by fire were not at first appreciated. And, in any event, the English were seldom sufficiently organized to mount a proper siege. Thus the value of fire against stone structures did not become apparent until the thirteenth century.

But the castle had arrived in English strategy, and any future campaign would have to reckon with it. In ensuing reigns events would be decided not so much by open battles, as at Senlac, but by dogged sieges of strongpoints.

English history has often seen a strong king followed by a weak son, but this was not true of William's successor. William II (Rufus) was nearly as able as his father, which was just as well for the stability of England.

As soon as the barons heard that the Conqueror had named his second son as heir to the English crown, and that the elder, the turbulent Robert, was assigned Normandy, they regarded this as an excellent opportunity for revolt. The pretext was that Robert, as the eldest, should have been king of England; but the object was undoubtedly self-interest, for the barons had decided that under the feckless Robert anarchy would come back and there would be no restraints on their power or excess.

The Red King countered this by a master-piece of diplomacy. He appealed to the English, promising that if they helped him put down the barons their own lives would be greatly improved as a reward. The English, with everything to gain and nothing to lose, responded magnificently. Their number was quoted at 30,000 but was doubtless considerably less. In any event they were sufficient to strike fear into the heart of the belligerent Bishop Odo of Bayeux, who garrisoned Rochester castle with 500 men-at-arms while himself prudently slipping off to Pevensey. Rochester was a useful strategic centre for it was well-placed for harassing London and Canterbury while securing its own supplies from the sea.

William Rufus proved a much better campaigner than the rebels expected. On this occasion he surrounded Rochester with a strong force, blocking every egress, and erected two powerful forts nearby. Roger of Mercia and a few others in the besieging army did their best to help the garrison but their treacherous activities were ineffectual.

Fortune favoured the besiegers in the shape of a plague of flies. These were not only exceptionally vicious, but also numerous enough to crawl over faces, eyes, food, and drink, without cease. Disease and discomfort—it was summer—was soon on such a scale that the garrison asked for terms. The King's reaction was unexpectedly violent. So far from meeting their requests he threatened to hang the entire garrison, and apparently meant it. His Norman supporters thought this was going too far, particularly as the castle contained a number of their relations. After much persuasion William relented sufficiently to spare their lives but confiscated all their possessions, including horses and arms. As the dejected garrison marched out William's English supporters shouted for halters, suggesting that the garrison should now be hanged. Odo was singled out for special invective, but William had made up his mind and all escaped death.

Professor E. A. Freeman in *The Reign of William Rufus* gives a full but slightly different account of William's campaign against Rochester of which the gist is as follows. The rebels held Tunbridge, Pevensey, and Rochester castles. William decided that the opening moves would be to capture Tunbridge and thereby cut off Pevensey from Rochester; this took a mere two days. However, Odo prudently used the two days to slip past William's forces and establish himself in Pevensey where he assumed Duke Robert would join him. The assumption proved ill-founded.

William, realizing that the capture of Odo was the essential objective promptly abandoned his move towards Rochester, marched to Pevensey, and began its siege. This was a tougher nut than Tunbridge; whereas the latter had been a mere mound, Pevensey was a Roman fortress strengthened by Norman builders. For six weeks Rufus assailed in vain, and the outcome of the rebellion was in some doubt. However, at long last Robert sent a fleet. This was intercepted by William's English forces who fought a lively battle on the beach which then separated the castle from the sea. The invaders lost the battle and most of their ships, and the failure of this relieving force meant that Pevensey castle was now on the verge of being starved out. Before this could occur Odo accepted fairly generous terms from Rufus; he was to supervise the surrender of Rochester castle and then leave the realm for ever.

Rufus appears to have behaved with uncustomary naïveté for he sent a vanguard with Odo assuming that the castle would be ready for his own occupation when he arrived. The defenders of Rochester had different plans, and when Odo appeared before the walls with a small force they made a swift sally and swept the Bishop and his bodyguard into the castle. As the defenders included such dubious characters as Count Eustace and Robert of Belesme, this act should have surprised no one; nor is it likely that Odo would have made any move to prevent it; probably he arranged it, as he soon became the life and soul of the defence.

William, as much incensed by his own stupidity as by this act of treachery, set about his task with great vigour. First, he raised even larger forces of English; secondly he built two temporary wooden forts to cut off the castle from any external relief, not that any was particularly likely as the rebels had no supporters nearer than Bristol and Durham. The rashness of the rebels' treachery was soon brought home to them by sickness and a plague of flies so that 'Nobody could eat unless his neighbour drove away the flies; so they wielded the flapper by turns'. Inevitably they asked for honourable terms; naturally these were at first refused. However, William was prevailed upon by his Norman supporters not to be too hard on the rebels, some of whose services he might find useful at a later date. After some delay he accepted the argument of expediency and spared the lives of the conquered. His English supporters were by no means pleased with this clemency and called 'Halters, bring halters, hang up the traitor bishop and his accomplices on the gibbet'. William, a man of his word, was deaf to their plea, and the worst that happened to the defenders was the loss of their English land. Odo was banished and does not appear again on the English scene but his supporters were soon back in favour, and most of them restored to their possessions.

Duke Robert's part in this rebellion was insignificant but his self-esteem was restored a little later in the rapid conquest of Ivry which had resisted his father for three years on a previous occasion. In this Robert was favoured by a hot summer which had dried the wooden roof of the hall. The besiegers shot red hot bolts on to this and captured the fortress in a day. The exploit is a little dimmed by the fact that the garrison numbered seven only.

One of the more colourful figures of the siege of Rochester was Robert de Rhuddlan. His normal activity was curbing the Welsh, and after Rochester he returned to this task with great vigour. Having strengthened Rhuddlan he pushed on and reinforced the old Roman Dictum as Diganwy castle. This stood on the opposite bank from the present Conway Castle. Cruel and avaricious though Robert was he did not lack personal courage and was eventually killed through taking on a Welsh force single-handed.

Having lost the first round against his brother, Duke Robert of Normandy proceeded to show how completely unfitted he was for any position of authority. His Norman barons did exactly as they pleased and the entire duchy was given over to chaos, vice, and murder.

In 1099, to everyone's surprise Duke Robert mortgaged his entire duchy to William Rufus for 10,000 marks (£6666). With this he equipped an army and set off on the First Crusade. The skill and gallantry he displayed in the campaign was completely at variance with his normal slothful and feckless behaviour. So great was his prestige that he was offered the

Kingdom of Jerusalem; he refused it and returned to his old ways. Unfortunately for him, he was far away when William Rufus was killed, and his youngest brother Henry was able to secure the throne for himself. Not unnaturally Robert attempted to change this arrangement but he was defeated in the battle of Tinchebrai (1106) and imprisoned in Cardiff Castle for the remaining thirty years of his life. Robert was said by some to have lived in comfort and luxury, though confined. Others give a different view, saying he had had his eyes put out (a favourite habit of Henry's) and that his clothes were mainly the King's cast-offs. However, as he lived to the age of eighty he was presumably adequately provided for.

A story, dating from January 1091, when Robert was besieging Courci, shows the peculiar conventions that were sometimes applied. 'The Duke laid siege to Courci, but unwilling to come to extremities with his great nobles, he took no measures for closely investing the besieged. Robert, however, used every resource of open attack and stratagem against the enemy for three weeks, employing various engines of war but was repulsed with shame. He caused a vast machine called a belfry to be erected over against the castles walls and filled it with all kinds of warlike instruments but even this failed of compelling the garrison to submit for as often as he began an assault on Courci, the powerful force from Grantmesnil hastened to the rescue and charging the assailants with fury drew them off from their intended attack. Meanwhile the garrison took prisoners William de Ferrers and William de Rupiere, whose ransoms were a great assistance to the besieged.' Presumably they bartered these valuable prisoners for fresh supplies.

'An oven had been built outside the fortifications between the castle gate and the assailants' belfry, and there the baker baked the bread required for the use of the garrison, because the siege was begun in such haste that they had no time to construct an oven in their new defences. It followed therefore that the thickest of the fight often raged round this oven, much blood was shed there, and many spirits departed by violence from the prison of the flesh.'

Seldom can cooking have been done in such distracting conditions!

'For the people of Courci stood in arms to defend their bread, while Belesme's followers tried to carry it off, so that many desperate conflicts occurred. It happened one day while the loaves were being baked in the oven, and the two hostile parties were engaged in a violent quarrel, the troops of both sides came up and a desperate conflict ensued, in which twenty men were killed, and more wounded, who never tasted the bread their blood had purchased. Meanwhile the friends of the besieged daily entered the castle in sight of the besiegers and, the duke taking no care to prevent it, conveyed to their comrades fresh supplies of arms and provisions to give them courage and support.'

On one occasion, Robert and his troops having been repulsed from the assault, those who pursued them made a squire mount into the belfry and set fire to it on the north side. The chronicler considers that this was God's will as the machine had been assembled on a holy day, i.e. Christmas.

The siege dragged on till the end of the month and then suddenly terminated on the news that William Rufus was arriving in Normandy with a large fleet. However that was not the last of the siege, for a neighbouring priest had a vision the day before the siege even began. He gave a full and detailed forecast of the subsequent fate in purgatory of the besiegers and many other Norman notables, including priests, lawyers, and women. The effect upon morale is not recorded.

In the event it turned out that William II was only marginally a better king than his brother might have been. He found promises easier to give than to remember, was oppressive and rapacious, was completely unprincipled, and had a furious temper. His court was undoubtedly the most debauched in English history. Homosexuals and harlots filled it; and did as they pleased. William of Malmesbury describes the scene: 'All military discipline being relaxed, the courtiers preyed upon the property of the country people, and consumed their substance, taking the very meat from the mouths of these wretched creatures. Then there was flowing hair and extravagant dress; and then was invented the fashion of shoes with curved points, then the model for young men was to rival women in delicacy of person, to mince their gait, to walk with loose gesture and half naked. Enervated and effeminate, they unwillingly remained what nature had made them; the assailers of others chastity, prodigal of their own. Troops of pathics and droves of harlots followed the court.' William II's achievements were not, however, limited to fields of greed and debauchery. Militarily he was as capable as his father. He pushed the English frontier up to Solway, and overran south Wales. North Wales proved harder to crack, and in spite of strong castles at Rhuddlan, Flint, and Montgomery he was unable to secure more than a loose control of the border regions. Attacks on Wales were mounted from three main bases: Shrewsbury, Chester, and Hereford, all of which had a screen of castles ahead of them.

Like his father, he was in the saddle when death came, but there was no lingering end for the Red King. Whether the arrow that killed him while he hunted in the New Forest was aimed deliberately, or struck him accidentally, will never now be known. Elaborate theories have been constructed to explain his death as the self-sacrifice of a hero-priest but in view of the fact that accidents were a regular occurrence (identical ones are recorded in the same period) and a mediaeval hunt, undisciplined as it was, must have been as dangerous to the hunter as the hunted, it seems pointless to look too deeply into the occurrence.

Ordericus tells three stories that illustrate the peculiar conventions of siege-warfare at the time. The first concerns the castle at Ballon which was held for Rufus by Belesme, and was being besieged by Count Fulk in 1098. The defenders decided to take the offensive and sent out spies disguised as beggars. These found out that the besiegers would all dine at the hour of tierce. A sally was organized, and surprised the besiegers in the middle of their meal. A hundred and forty knights and a crowd of foot-soldiers were taken prisoner. When Rufus arrived on the scene he accepted their parole and ordered a meal to be prepared for his captives. When it was suggested to him that the prisoners might break their parole he rejected the idea with scorn.

Equally extraordinary is William's siege of Chaumonton. For some unexplained reason the defenders aimed their arrows at the horses and not at the men. In spite of this self-imposed handicap their resistance was successful and William called off the siege.

At the siege of Mayet Rufus postponed the attack from Saturday morning till Monday. This was unexpected in view of his known impiety. The defenders used the respite to put wicker crates along the walls to break the force of the stones hurled from the mangonels. A wide ditch made the task of the besiegers difficult so the king had it filled up with horses and mules. It is said that Belesme also added villeins. However the filling was probably mostly wood because burning charcoal was thrown down by the besieged in order to set it alight. William was nearly killed by a stone which crushed the man next to him, on which the garrison jeered 'ho the King now hath fresh meat; let it be taken to the kitchen and made ready for his supper'. Eventually he was persuaded to leave by the strange argument that the defenders had an advantage over the attackers.

There is also a note on the castle at Ivri which shows the dangers of being too good a designer—and too domineering a wife.

'This is the famous castle of great size and strongly fortified which was built by Alberede wife of Ralph, count of Bayeux. It is said that Alberede, having completed this fortress with vast labour expense, caused Lanfred, whose character as an architect transcended that of all the other French architects of that time, and who after building the castle at Pithiviers was appointed master of these works, to be beheaded that he might not erect a similar fortress anywhere else. She also was put to death by her husband, on account of this same castle of Ivri, having attempted to expel him from it' (Ordericus).

Mediaeval customs and philosophy are illustrated by the case of Humphrey Horenc. His eldest son was named Havise, which was also his wife's name. 'He gave the herbage of the whole vill free from commonage, and all the land in the parish, whether in grass or tillage, to be cultivated by the tenants there, reserving only the Champarty (a form of ground rent). Not long afterwards he was by God's providence affected with a painful disease in his privy-parts and having the fear of death before his eyes became a monk in the abbey of Bec. His son built a very strong castle at Breval and filled it with fierce freebooters who ruined numbers. He surprised the castle of Ivri by a skilful stratagem, defeating and making prisoner

William de Breteuil its master whom he threw into close confinement. For his ransom he extorted violently a thousand livres of Dreux and the stronghold of Ivri, taking to wife his daughter Isabel by whom he had seven sons.'

The monk chronicling these events appears to have no reservations about the good fortune of the father or the piety of the son; nor any comment to make on the situation of William of Breteuil except that he 'suffered much during the ensuing heat so that for his sins he was compelled to endure the rigours of that pestilential season'.

However certain people and practices do call down criticism. One is Fulk, Count of Anjou. 'His feet being very deformed he had shoes made of unusual length, and very sharp at the toes, so that they might conceal the excrescences, commonly called bunnions which caused his feet to be so ill-shaped. This new fashion became common throughout the west and wonderfully pleased light-minded people and the lovers of novelty. . . . But now men of the world sought in their pride patterns of dress which accorded with their perverse habits, and what formerly honourable people thought a mark of disgrace, and rejected as infamous, the men of this age find to be sweet as honey to their taste, and parade on their persons as a special distinction.' At this time effeminacy was the prevailing vice throughout the world.

'The habits of illustrious men were disregarded, the admonitions of priests derided, and the customs of barbarians adopted in dress and in the mode of life. They parted their hair from the crown of the head on each side of the forehead, and let their locks grow long like women, and wore long shirts, and tunics closely tied with points. . . . Our wanton youths are sunk in effeminacy, and the courtiers study to make themselves agreeable to women by every sort of lasciviousness. . . . Sweeping the dusty ground with the prodigious trains of their robes and mantles they cover their hands with gloves too long and too wide for doing anything useful, and encumbered with these superfluities lose the free use of their limbs for active employment. The fore-part of their heads is bare after the manner of thieves, while on the back they nourish long hair like harlots.'

Debauched, effeminate, and ludicrous though these men appeared to the chroniclers, they were nevertheless capable of engaging in desperate and bloody warfare. 'The clang of arms gave token of frequent conflicts, and the soil was watered with the blood of the slain.'

The honour of being first to strike a blow, to scale the battlements or to cross the threshold was eagerly sought. The story of Taillefer the Norman jester who rode first into the fray at the Battle of Hastings and was killed first is too well known to need repetition here. Later the distinction of striking the first blow depended on rank, and an over-ardent but low-born fighter would be as likely to be cut down by his own side as the opposition. The top of the scaling ladder was however freely available to anyone who wanted to make use of it.

At Marrah (December 1098) it was stated, 'Scaling ladders were raised against the walls but the violent outcries and threatening demonstrations of the Turks discouraged everyone from venturing to mount them. The citizens indeed thought that their resistance on the present occasion would be as successful as it had been against Raymond Pilet but Count Raymond caused a machine to be built of wood, to run on four wheels that it might easily be moved. It was so lofty that it commanded the walls and reached to the battlements of the towers. The structure was rolled forward against one of the towers, the trumpets and clarions sounded, and the troops under arms invested the whole circuit of the walls, the crossbowmen and archers discharged their bolts, and the party in the wooden tower hurled below immense stones, while the priests and clerks offered earnest prayers to the Lord for his people. William of Montpellier and many others fought from the one machine, overwhelming the citizens beneath with stones and darts, and easily killed them by crushing their shields, helmets and heads; others made incessant attacks on the defenders of the walls with iron hooks. On the other hand the Turks directed their arrows and missiles against the Christians from the towers, they also threw Greek fire into the machine, and left nothing untried. The struggle was prolonged till evening. At length Gouffier de Tours, a knight of the Limousin, of high birth and extraordinary daring, was the first to mount the scaling ladder and reached the top of the wall. Some soldiers, few in number, ascended after him, for the ladder was broken and fell in pieces. Gouffier, however, held his footing on the battlements manfully, driving back the pagans, and at the same time calling his comrades to his aid, both by gestures and voice. They soon raised another scaling ladder, by means of which so many knights and soldiers mounted to the top that they occupied a long extent of the wall, from which they entirely drove the garrison. The Infidels now rallied, and renewed the attacks with so much determination, sometimes charging the Franks with such impetuosity that some of them in their terror leaped from the wall. However a strong body maintained their ground, and resisted the determined attacks of the enemy until the Christians below had undermined the wall and made a breach for the besiegers to enter. . . . The Christians spread themselves through the conquered place, and mercilessly pillaged it of all the wealth they could find in the houses and cellars, giving no quarter to the Saracens but putting them almost all to the sword.' If one looks for a reason for the desperation of most mediaeval fighting it is in this account of a Christian victory; the fate of the victims of paganism could not easily have been worse but was unlikely to have been better.

The scaling ladder was the only means of making an individual bid for victory; every other means of assault meant teamwork. At Jerusalem 'the crusaders made a vigorous assault on the city, and it was believed they would have taken it if they had been sufficiently supplied with scaling ladders. They made a breach in the outer wall, and raised one ladder against the inner one. The Christian knights mounted it by turns and fought with the Saracens on the battlements hand to hand with swords and lances. In these assaults many fell on both sides but most on the side of the Gentiles. The trumpets sounding the recall the Christians at length withdrew from the combat, and returned to their camps.'

In another version of the same incident there is an account of a keen rivalry to mount the 'one ladder' mentioned above. The honour fell to Raimbaud Creton whose hand was cut off the moment he reached the top and grasped the battlement.

Having broken off the siege the Crusaders endured agonies of hunger and thirst. The army chiefs said, 'We are in difficulties on all sides, bread is wanting; the water has failed. We ourselves are, in fact, closely blockaded while we fancy we are besieging this city. We can hardly venture outside our camp, and when we do, return empty.'

Desperation drove the Crusaders to a further attack. Both sides worked day and night to build defences and assault towers. 'On Saturday Duke Godfrey's machine was transported in the dead of night to the foot of the walls, and erected before sunrise. The Count of Tholouse caused his machine, which might be called a castle of wood, to be placed near the wall on the south of the place, but a deep hollow prevented it being joined to the wall. Such machines cannot be guided on declivities nor carried up steep places, and can only be transported on level ground. Proclamation was therefore made through the camp that whoever should cast three stones into the hole should for so doing receive a penny. In consequence all the people who were weary of delay lent a hand willingly to the proposed work.'

Presumably the stones were heavy and hard to find and the Saracens used the carriers for target practice. The hole took three days to fill and the general exertions tired the assault troops so much they were given the week-end to recuperate.

The battle was rejoined on July 15th, 1099, and once more there was intense rivalry to be the first to cross the walls of the Holy City. The first two soldiers to reach the top of the walls were Belgians who used scaling ladders, the third was a Norman who worked his way over astride a piece of wood extending from the tower to the battlements. The Christians, full of righteous indignation about the way the Holy Places had been misused and profaned, proceeded to slaughter the inhabitants wholesale, 'No one knows the number of the slain but the floor of the temple was knee deep in blood, and great heaps of corpses were piled up in all quarters of the city, as the victors spared neither age, sex, rank nor condition of any kind.'

Ordericus recounts the sequel to the fall of Jerusalem without criticism but without enthusiasm. It seems unlikely that the slaughter was the result of religious indignation at the occupation of the Holy City; it was more probably the pent-up fury of men who had been goaded by dust, desert winds, thirst, hunger, and frustration. Women, or even children, were unlikely to be exempt from paying the penalty of being in a centre of resistance. Mention is often made of their deeds on the walls where they were fully capable of dropping stones or pouring incendiary material. An account of the same siege by William of Malmesbury is given in the next chapter, and highlights different activities.

But the savagery of mediaeval warfare did not go without criticism. In the higher ethics of the age all fighting was thought to be wrong, that between Christians and heathens exculpable, but between Christian and Christian to be morally indefensible. The fact that knights spent most of their time defying ecclesiastical rules probably accounted for their remarkable piety when their lives appeared likely to be terminated abruptly, for one reason or another. Usually that piety was directed to endowing a church or ancillary activity. Monks appear to have had a keen business sense. When William de Warenne, 1st Earl of Surrey, died at Castle Acre in 1088 the Abbot of Ely heard the devil crying for his soul. The Abbot was disposed to believe his ears because some years before Warenne had received a substantial piece of abbey lands from the Conqueror, and requests for him to restore them had been unavailing. The monks reported that he had left them 100 shillings and that this had been sent immediately by the late Earl's wife. The story seems to have overlooked the fact that she had predeceased him by three years. The dispute about the land was still not settled a hundred years later. Warenne was buried in the Priory of St Pancras which he had established at Lewes. The deduction to be drawn from this and other similar occurrences was that mediaeval knights and barons were very generous to those with whom they were identified and whose spiritual support was considered reliable, but rather less dutiful to others; on occasion they would despoil other monks of their property, relying on the prayers of their own clerical supporters to outweigh the curses of the dispossessed.

❋ 4 ❋

The Castle as an Instrument of Government

Henry I (1100-1135)

HENRY I was the youngest and probably the ablest of the Conqueror's sons. William I, who left him no land and only a small sum of money, said in explanation: 'He will soon lack neither'.

The sudden death of William Rufus, and the fact that his other brother was away on the crusade, gave Henry his opportunity. Realizing that the northern barons would prefer erratic brother Robert as their master, Henry decided to forestall them. Within three days he was proclaimed king and crowned. News travelled slowly in 1100 and the northerners did not hear of Rufus's death until Henry had installed himself firmly on the vacant throne. Once there he made such a show of redressing wrongs, as well as restoring lost property and positions, that both the English and Normans welcomed him gladly.

But there were, as the King realized, dissentient voices, and the loudest of these was that of Robert of Belesme, Earl of Shrewsbury. Belesme had a reputation for cruelty which shows he was psychopathic (his mother was also a sadist). His deep cunning and evil intentions were masked by an air of charm and affability, and the sweeter the manner the more devilish the ensuing crime. Malmesbury reports him as having scratched out the eyes of his godchild in order to annoy the father.

The rebellion lasted until the summer of 1101, and turned upon the sieges of Arundel and Bridgenorth. Little of Bridgenorth castle now remains but its commanding position, about 200 feet above the river, and the difficulties of assault are obvious. Two sides of the triangle are made impregnable by siting, on the third the man-made defences were formidable, and a marshy valley separated it from Oldbury. Belesme, though a bloodthirsty tyrant, was an expert at castle building, being responsible for the siting and planning of Gisors, one of the most formidable castles in Normandy. In all probability Bridgenorth was not planned when Belesme heard the news of Henry's accession but at this and other of his strongholds it is said that the work proceeded by night as well as by day. In spite of these desperate preparations Henry captured the stronghold after a siege of only three months. The key to his success was the malvoisin, the wooden tower that enabled him to make the life of the defenders so unbearable that they surrendered. Belesme was exiled but was captured in a later rebellion; he spent the rest of his life in prison at Wareham Castle.

The Anglo-Saxon Chronicle contains the following tactical information: 'And the King went and besieged the castle of Arundel. When however he could not take it by force so quickly he had castles made before it, and garrisoned with his men, and then with all his army marched to Bridgenorth and stayed there until he had the castle.'

Unfortunately no time is given for the construction of these emergency siege castles, otherwise known as malvoisins, but their purpose is interesting. With a small holding force he was able to prevent supplies reaching Arundel castle and could leave knowing that the situation would be much the same on his return, as the inhabitants of Arundel would not be able to drive off the containing force.

William of Malmesbury gives a slightly different account. 'The following year Robert de Belesme rebelled, fortifying the castles of Bridgenorth and Arundel against the King, carrying thither corn from all the district round Shrewsbury, and every necessary which war requires. The castle of Shrewsbury, too, joined the rebellion, the Welsh being inclined to evil on every occasion. In consequence the King, firm in mind and bearing down every adverse circumstance by valour, collecting an army, laid siege to Bridgenorth, from whence Robert had already retired to Arundel presuming from the plenty of provision and the courage of the soldiers that the place was abundantly secure. But after a few days the townsmen, impelled by remorse of conscience, and by the bravery of the King's army, surrendered; on learning which Arundel suppressed its insolence, putting itself under the King's protection, with this remarkable condition, that its lord, without personal injury, should be suffered to retire to Normandy. Moreover the people of Shrewsbury sent the keys of the castle to the King as tokens of present submission and pledges of their future obedience.'

The sudden collapse of resistance at Bridgenorth probably had several causes: Belesme was a hated tyrant; the absence of the feudal lord would have a bad effect on morale; if the local people who had been brought in for the siege decided to surrender the castle was doomed, whether or not the loyal troops decided to slaughter the defeatists.

Belesme, although an expert on castle design, was an incompetent commander, and was defeated more often than not when he took the field.

Ordericus gives a further, and fuller, version of Bridgenorth. 'The King of England did not, like his brother, abandon himself to sloth, but in the autumn arranged the military forces of the whole of England, and leading them into Mercia, besieged Bridgenorth for three months. Robert de Belesme had retired to Shrewsbury, entrusting Bridgenorth to Robert, son of Corbet, with eighty stipendiary men at arms. He had now entered into an alliance with the Welsh, frequently employing the troops to attack the royal army. He had disinherited William Pantoul, a brave and experienced knight, when he proferred his valuable services at a time they were urgently needed. Being thus rejected with disdain William Pantoul went over to the king who received him graciously. He gave him the command of two hundred men, and entrusted to him the custody of Stafford castle in the same neighbourhood. This knight proved Robert de Belesme's worst enemy, never ceasing from persecuting him both by his counsels and his arms till his ruin was completed.'

However Pantoul's zeal was not matched by the other Norman earls and barons who were afraid that if Belesme were crushed Henry would take the opportunity of reducing baronial power generally. Accordingly they went to the King in a body and tried to soften his resolution to crush Belesme. Unfortunately for them a body of some three thousand troops became aware of what was going on and shouted 'Henry trust not these traitors. They are endeavouring to deceive you and prevent the exercise of your royal justice.'

Henry listened to the troops, and dismissed the barons. He then bought off the Welsh princes and added their forces to his own. 'He also sent for three of the principal townsmen, and swore to them publicly that unless the place was surrendered to him within three days he would hang all of them he could lay hands on. Apparently the citizens had been put in the castle from the neighbouring town of Quatford. Pantoul was brought into the subsequent negotiations, and as a result they sent a message to Belesme telling him they would now surrender. The stipendiary troops, i.e. Belesme's personal troops, were kept in ignorance of these proceedings but when they discovered the trend of events flew to arms. However they were 'blockaded in one part of the fortress'.

Belesme was in some distress but was still strongly established in Shrewsbury. 'The king now issued orders for his army to march by the Huvel Hegeve [Evil Hedge] and lay siege to Shrewsbury, which stands on rising ground washed on three sides by the river Severn. The road was for a thousand paces full of holes, and the surface rough with large stones and so narrow that two men on horseback could scarcely pass each other. It was overshadowed on both sides by a thick wood in which bowmen were placed in ambush ready to inflict sudden wounds with hissing bolts and arrows on the troops on their march. There were more than sixty thousand* infantry in the expedition, and the King gave orders that they should clear a broad track by cutting down the wood with axes so that a road might be made for his own passage and a public highway for ever afterwards. The royal command was promptly performed and vast numbers of men being employed the wood was felled and a very broad road levelled through it. . . .'

This last was too much for Belesme who promptly surrendered and was banished from the realm.

Although there were numerous castles in England by this time Henry was by no means satisfied, and pressed on with siting, planning and construction. 'Castella erant crebra per

* Doubtless a wild exaggeration; probably 16,000. Numbers are often unreliable, usually exaggerated though sometimes diminished.

totam Angliam.' Most of the original makeshift motte and bailey constructions had now been replaced by formidable stone buildings. All through the west country up to Wales powerful castles were completed. The chain extended from Cardiff through Sherborne, Devizes, and Newark. Other castles such as Lewes and Reigate in the south and Coningsborough in the north were held by loyal barons such as William de Warenne (although he slipped from grace for a short period). Numbers of castles fell into Henry's hands after the Baron's revolt, and they remained royal possessions. Bridgenorth, Shrewsbury, and Arundel came from Belesme, Hinckley from Grantmaisnel, and Durham from Flambard. Later, others were added to these. The most powerful addition of all was probably Kenilworth, where considerable ingenuity was shown in the use of water defences—perhaps with memories of Hereward in mind.

The sieges in which Duke Robert distinguished himself on the First Crusade are dramatically described by William of Malmesbury. At Antioch, 'And now everything which could be procured for food being destroyed around the city, a sudden famine, which usually makes even fortresses give way, began to oppress the army, so much so that some persons seized the pods of beans before they were ripe, others passed parboiled thistles through their bleeding jaws into their stomachs. Others sold mice, or such like dainties, to those who required them, content to suffer hunger themselves so that they could procure money. Some too there were who fed their corpse-like bodies with other corpses, eating human flesh.' Unfortunately such practices are always likely to return under certain conditions, and were not unknown in the Far East during the Second World War.

William of Malmesbury relates a story which shows the peculiar attitudes then prevailing. 'To revenge this disgrace the Turks wreaked their indignation on the Syrian and Armenian inhabitants of the city, throwing by means of their balistae and petraries, the heads of those whom they had slain into the camp of the Franks, that by such means they might wound their feelings.'

The English castle and fortified town, adequate though they were for home requirements, were crude and unsophisticated compared with their middle-eastern counterparts. It is not, therefore, surprising that the Crusaders, having had to pit themselves against much more subtle defences than they had previously experienced, should endeavour to embody some of the lessons they had learnt in their own constructions. Duke Robert of Normandy was present at the attack on the Tower of David, a fortress that defended Jerusalem on the west, and which was built of square stone blocks soldered with lead. 'As they saw therefore that the city was difficult to carry on account of the steep precipices, the strength of the walls, and the fierceness of the enemy, they ordered engines to be constructed. But before that on the seventh day of the siege, they had tried their fortune by erecting ladders, and hurling swift arrows against their opponents, but as the ladders were few, and perilous to those who had to mount them, since they were exposed on all sides, and nowhere protected from wounds, they changed their design. There was one engine which we call the Sow, the ancients Vinea, because the machine, which is constructed of stout timbers, the roof covered with boards and wickerwork, and the sides defended with undressed hides, protects those who are within it, who, after the manner of a sow, proceed to undermine the foundations of the walls. There was another, which, for want of timber, was but a moderate sized tower, constructed with several floors, one above the other, all of which contained soldiers; this was intended to equal the walls in height. And now the 14th day of July arrived, when some began to undermine the wall with the sows, others to move forward the tower. To do this more conveniently they took it forward towards the works in separate pieces, and, putting it together at such distance as to be out of reach of bowshot, advanced it on wheels nearly close to the wall. In the meantime, the slingers with stones, the archers with arrows, and the crossbow-men with bolts, each intent on his own department, began to press forward, and dislodge their opponents from the ramparts; soldiers, too, unmatched in courage, ascended the tower, waging nearly equal war against the enemy with missile weapons, and with stones. Nor indeed were our foes at all remiss; but trusting their whole security to their valour, they poured down grease and burning oil upon the tower, and slung stones on the soldiers, rejoicing in the completion of their desires by the destruction of multitudes. During the whole of that day the battle was such that neither party seemed to think they had been worsted; on the following, which was the 15th July, the business was decided. For the Franks, becoming more experienced from the event of the attack of the preceding day, threw faggots flaming with oil on a tower adjoining the wall, and on the party who defended it, which, blazing by the actions of the wind, first seized the timber, and then the stones, and drove off the garrison. Moreover the beams which the Turks had left hanging down from the walls in order that, being forcibly drawn back they might by their recoil, batter the tower in pieces in case it should advance too near, were by the Franks dragged to them, by cutting away the ropes, and being placed from the engine to the wall, and covered with hurdles they formed a bridge of communication from the ramparts to the tower. Thus what the infidels had contrived for their defence became the means of their destruction; for then the enemy, dismayed by the smoking masses of flame and by the courage of our soldiers, began to give way. These, advancing on the wall, and thence into the city manifested the excess of their joy by the strenuousness of their exertions' (Malmesbury).

After sieges like this there was usually a wholesale massacre and this was no exception. The knowledge that surrender meant slavery, and defeat led to massacre, encouraged besieged garrisons to fight with the fury of men whose only hope lies in victory.

Siege warfare placed a high premium on personal courage and initiative. The fall of Château Gaillard, as will be seen later, is a striking example of how one man's action can lead to victory. Doubtless there were numerous occasions when courage and initiative resulted in a swift end for the possessor but the difference between mediaeval and modern warfare is that in the former individual action was more important. William of Malmesbury quotes that at the siege of Rome Godfrey of Bouillon was 'the first to break through that part of the wall which was assigned for his attack, and facilitated the entrance of the besiegers. Being in extreme perspiration, and panting with heat, he entered a subterraneous vault which he found in his way, and when he had there appeased the violence of his thirst by an excessive draught of wine, he brought on a quartan fever.'

The results of this, or the general rigours of the campaign, stayed with Godfrey for some time but at Antioch he was his old self again. 'There, with a Lorrainian sword he cut asunder a Turk who had demanded single combat, and one half of the man lay panting on the ground while the horse at full speed carried away the other; so firmly the miscreant sat.' On another occasion he killed a ferocious lion with spear and sword. Perhaps his most notable feat was when he was attacked by a Turk. 'He clave him asunder from the neck to the groin, by taking aim at his head with his sword, nor did the dreadful stroke stop here but cut entirely through the saddle and the backbone of the horse.'

Godfrey was undoubtedly a formidable figure but his brother Baldwin eclipsed him. Of Baldwin it was said 'he fell little short of the best soldier that ever existed'. Baldwin took Azotus in three days but previously Godfrey had failed in the same situation. A hazard of the siege tower was that it would sometimes sink into loose ground, prepared for the purpose, and topple sideways. This misfortune had occurred to Godfrey. 'For indeed, when by means of scaling ladders he had advanced his forces on the walls and they, now nearly victorious had gotten possession of the parapet, the sudden fall of a wooden tower, which stood close to the outside of the wall, deprived them of the victory and killed many, while still more were taken and butchered by the cruelty of the Saracens.'

With this fresh in his mind Baldwin set about his task with commendable thoroughness: 'He ordered engines to be constructed. Petraries [stone-slinging machines] were therefore made, and a great tower built of twenty cubits in height, surpassing the altitude of the wall.'

A cubit is not a precise measurement, being based on the approximate length of a man's forearm, but was not less than eighteen inches nor more than twenty-two. An approximate estimate of the height of this tower is 35 feet. Later these towers would need to be double the height to have any effect on castle walls. However the besiegers could not wait for the completion of this and 'impatient of delay and such lingering expectation, erecting their ladders and attempting to overtop the wall arrived at the summit by the energy of their efforts, with conscious valour indignantly raging, that they had now been occupied in conflict with the Saracens fifteen days and had lost the whole of that time; and although the Caesareans resisted with extreme courage, and rolled down large stones on them as they ascended, yet despising all danger, they broke their opponents in a close body, and fought with outstretched arm and a drawn sword. The Turks, unable to sustain the attack and taking to flight, either cast themselves headlong, or fell by the hand of their enemies. Many were reserved for slavery; a few for ransom.'

An odd feature of the end of this and other sieges was that the Turks hid their money in their mouths; this custom was known to the Christians who hit them on the neck to make them disgorge. 'The scene was enough to excite laughter in a bystander to see a Turk disgorging bezants when struck on the neck by the fist of a Christian.'

Henry I, although a successful campaigner, was not an enthusiast for personal conflict and exposure to danger. He was not necessarily less brave than other kings of the period—and indeed had been struck on the helmet by an arrow during an expedition to Wales—but his approach was different. They enjoyed the heat and dust of conflict; he preferred to direct the battle from outside.

Fashion seemed to dominate life then as much as it does to-day. The chronicler reports 'a circumstance occurred in England which may seem surprising to our long-haired gallants, who forgetting what they were born, transform themselves into the fashion of females by the length of their locks. A certain English knight, who prided himself on the luxuriancy of his tresses, being stung by conscience on the subject, seemed to feel in a dream as though some person strangled him with his ringlets. Awakening in a fright, he immediately cut off all his superfluous hair. The example spread throughout England, and as recent punishment is apt to affect the mind, almost all military men allowed their hair to be cropped in a proper manner, without reluctance. But this decency was not of long continuance; for scarcely had a year expired ere all who thought themselves courtly, relapsed into their former vice; they vied with women in the length of their locks, and wherever they were defective, put on false locks.'

Henry, described as being peace-loving and pious, established such a grip on the country that he was able to spend the last three years of his reign in France without serious trouble in his kingdom.

During his reign the network of fortresses planned by William I was completed and consolidated. The temporary buildings of the past were mainly replaced by stone works, most of them strong through their massive construction rather than subtlety of design. But the First Crusade began to exercise considerable influence. The Middle East was far advanced in fortress and castle construction, as already noted, and as early as 2000 B.C. fortresses had existed in the Middle East which were not surpassed until the great castle-building period of Edward I.

Nevertheless Henry's castles showed some notable advances. Previously defence had taken place from the battlements. Henry introduced air-vents lower down in the walls, from which the archers could still operate when a section of the battlement was temporarily unserviceable through heavy attack. Another innovation was the portcullis which could trap the vanguard of an invading force, a much-feared device that was difficult to counteract. However, if sufficient props were placed underneath to make it unserviceable the vital entrance would be only partially blocked.

A factor which must undoubtedly have had its effect on sieges and battles was the general wretchedness of conditions of life throughout the reign. In 1103 the Anglo-Chronicle states: 'This was a very grievous year in the country through all sorts of taxes, and cattle plague and ruin of crops—both corn and all the produce of trees. Also, on the morning of 10th August the wind did so much damage to all crops in this country that no one remembered it ever doing so much before.'

In 1104, 'It is not easy to describe the miseries the country was suffering at this time, because of various and different injustices and taxes that never ceased or diminished, and always wherever the King went there was complete ravaging of his wretched people caused by his court, and in the course of it there were burnings and killings'.

The following year, 'this was a very grievous year because of the ruin of crops and the various taxes that never ceased'.

Four years later (1109), 'there were many thunderstorms and very terrible they were'. And again in 1110, 'This was a very severe year because of storms by which the products of the soil were badly damaged and the fruits of trees over all this country nearly all perished'.

In the following year, 'This was a very long and troublesome and severe winter, and as a result all the produce of the soil was very badly damaged, and there was the greatest cattle plague that anyone could remember.'

1112 started well, 'A very good year and very productive in woods and fields but it was very troublesome and sorrowful because of excessive plague'. The year 1113 passed without great distress but 1114 had more than its share of violent October winds. A curiosity of the year was an ebb in the Thames—a much broader river then than now—so that it could be crossed easily on foot on the 10th of October. It was caused by freak winds affecting the tide.

In 1115 the winter was 'so severe, that with snow and frost, that nobody then living remembered any more severe, and because of that there was excessive plague among cattle'.

The following year the winter was also 'bad and severe' but the main hardship was caused by 'the excessive rains' that came shortly before August and were causing much distress and toil when Candlemas (2nd February) was reached.

1117 saw excessive storms, and was 'a disastrous year for corn because of the rains that hardly ceased nearly all the year'.

1118 produced thunder, lightning, and excessive gales, and was followed by an earthquake that was especially severe in Gloucestershire and Worcestershire.

Another damaging earthquake, centred on Somerset and Gloucestershire, occurred in 1122. In 1124 the weather was so bad that corn prices reached unprecedented heights. 1125 brought floods, famine, and disease.

In brief, nearly half the years of Henry's reign brought natural disasters of one sort or another. Such conditions mean short sieges for they prevent the collection of adequate food supplies.

The achievement of the Norman kings can only be fully appreciated if one understands the potential power against them. Robert de Belesme is one example. Another but different example of personal influence was Robert, Earl of Mellent, 'and his advice was as if the oracle of God had been consulted. He possessed such mighty influence in England, as to change by his single example the long established modes of dress and diet. Finally the custom of one meal a day is observed in the palaces of all the nobility through his means. He is blamed, as having done, and taught others to do this more through want of liberality than any fear of surfeit or indigestion, but undeservedly' (Ordericus). Henry of Huntingdon, while admiring Robert, thinks that the fashion was frequently followed through meanness.

One of the most curious sieges of mediaeval times occurred when Henry quarrelled with his son-in-law Eustace de Breteuil. Eustace, who had married Juliana, one of the King's illegitimate daughters, was himself the illegitimate offspring of a turbulent family which had on numerous occasions engaged in rebellion, and had become infamous when one of its members got rid of a feudal rival by sending him a present of gloves and a helmet; the recipient died in great agony for both were poisoned.

Eustace, egged on by kinsmen and cronies, demanded the return of the castle of Ivri, which had been forfeited after their last unsuccessful rebellion. It had belonged to his family for several generations and, although other members had a stronger claim, no one had a better chance of obtaining it than Eustace. In the past he had proved himself a redoubtable warrior, and was regarded as a useful and, for the times, reliable ally. Henry was reluctant to relinquish such a key strong-point but he did not want to quarrel with Eustace whose friendship was valuable. He therefore played for time by saying he had the matter under consideration, and as a mark of good faith handed over the son of the custodian of the fortress, Ralph Harenc. In return Eustace handed over his two daughters to Henry (who were, of course, the King's granddaughters).

The arrangement did not suit Eustace's friends who were anxious to see Eustace start hostilities from which there might be chances of plunder. They therefore goaded Eustace into blinding the boy, who was then sent back to his father.

Harenc promptly sought redress from Henry who handed over his granddaughters to the enraged father, with full permission to do what he wished. Harenc blinded the girls and cut off the tips of their noses. Then, loaded with presents from Henry, he returned to Ivri and sent a message to Eustace telling him he could collect his daughters.

Eustace promptly declared war on Henry, and put all his own castles on a defensive footing. He presided over four himself and sent his wife, Juliana, with a substantial garrison to hold the fifth, which was the castle of Breteuil.

The townspeople of Breteuil, not wishing to be involved in hostilities against Henry, did not back Juliana, so the unfortunate mother, lacking the benefit of local support, asked for a parley. When the King appeared she launched a crossbow bolt at him, but missed. Henry retaliated by ordering the castle drawbridge to be broken, and Juliana, a prisoner in her own castle, agreed to surrender. Henry, however, would not allow her to leave. She thereupon 'let herself down from the summit of the walls without support, and as there was no bridge she descended into the foss indecently with naked legs. This took place when the castle ditch was full of snow water which being half frozen her tender limbs of course suffered in her fall from the severity of the cold.'

In spite of these difficulties Juliana escaped and, at a later date when all hope of revenge was gone, was reconciled to Henry.

Although the act of an infuriated mother can scarcely be described as treachery—although it was so labelled by the chroniclers of the time—Henry was very familiar with that activity. A case in point occurred at Pontaudemer in 1123. 'Meanwhile the king was besieging an enemy's castle but had suspicions of many of those who, admitted to familiar intercourse with him, loaded him with flatteries; and discovering their perfidy, he considered them as disloyal men. However the king reduced to ashes the town, which was of great size and very rich, and sharply assaulted the castle. He himself carefully looked to everything, running about like a young soldier, and animated all with great spirit to perform their duties. He taught the carpenters how to construct a berfrey, jocularly chid the workmen who made mistakes, and encouraged by his praise those who did well to greater exertions. [The berfrey in question was 24 feet higher than the walls.] At last he completed his machines, and by frequent assaults on the besiegers, which occasioned them serious loss, compelled them to surrender the place.'

According to Simeon of Durham this siege lasted seven weeks. During this time Robert de Chandos, the warden of Gisors, was invited to a parley in a citizen's house, while the whole town was filled with soldiers disguised as peasants coming to the market. Fortunately for Chandos he was delayed by a long domestic discussion with his wife, and the trap was sprung before he was in it.

Had the siege of Pontaudemer lasted a little longer Henry would have been in an awkward situation, but in the event he was able to move swiftly to Gisors. Such incidents undoubtedly contributed to the suspicion and ruthlessness which characterised his rule. However, the uncritical Ordericus cannot but acknowledge in his account of the Normans, 'Heathens as they were, the Roman legions committed no such crimes.'

An early experiment in physiology was conducted in 1119 by Siward, King of Norway, who journeyed to Constantinople. 'His men dying in numbers in this city he discovered a remedy for the disorder by making the survivors drink wine more sparingly, and diluted with water; and this with singular sagacity; for pouring wine on the liver of a hog, and finding it presently dissolved by the acridity of the liquor, he immediately conjectured that the same process took place in men, and after dissecting a dead body he had ocular proof of it.'

One of the most extraordinary incidents occurred in 1119. Louis of France had despatched a force into Normandy. Ralph de Guader opposed it, 'engaging them with vigour and causing them lamentable losses by the fierce blows dealt on them with lance and sword. He caused all the gates of the castle to be thrown open at their approach but no one ventured to force his way through the open doors, the astonishing courage of their opponents sufficiently repelling them. The battle raged furiously outside the three gates, and brave warriors fell in great numbers on both sides' (Ordericus). The flamboyance of this tactical move succeeded. The French decided to stay and fight rather than try to by-pass the castle; in consequence they were delayed and lost their strategic advantage to the English.

A reason why men valued their lives cheaply occurs in the Anglo-Saxon Chronicle. The entry for the year 1124 records.

'In the course of this same year after St Andrew's Day, before Christmas, Ralph Basset and the King's thegns held council at Hundebroye in Leicestershire and hanged there more thieves than had ever been hanged before; that was in all forty-four men in that little time; and six men were blinded and castrated. A large number of trustworthy men said that many were destroyed very unjustly but Our Lord God Almighty that sees and knows all secrets—He sees the wretched people are treated with complete injustice: first they are robbed of their property and then they are killed. It was a very troublous year: the man who had any property was deprived of it by severe taxes and severe courts; the man who had none died of hunger.' In the following year all moneylenders were summoned to Winchester. When they got there they were taken one by one and each deprived of the right hand and castrated. The King's justice was, indeed, as harsh as the conditions under which his subjects had to live but his authority was maintained.

When Henry died the Chronicle reported. 'He was a good man and people were in great awe of him. No one dared injure another in his time. He made peace for man and beast. Whoever carried his burden of gold and silver nobody dared say anything but good to him.'

Henry also built eleven castles in Normandy. The foolhardy Louis VI of France had invaded the Duchy and Henry decided to secure it from further trouble. All were typical rectangular Norman designs. The ground floor was a store room and could only be reached from the interior by a ladder or wooden stair. Subsequently this part of the castle would become the hall. The first floor was the quarters of the owner or castellan (castle-keeper). Staircases were hardly developed and where they existed were of wood. Defence was conducted from the battlements, and where slits existed in the walls they were for ventilation and light, not archery. As the entrance to the keep was high above ground level it was reached by a footbridge that could easily be removed in an emergency. The keep at Kenilworth gives a very good idea of the external appearance of a castle of that date. A few experiments of polygonal castles had been made in order to overcome the 'blind-spot' corners of the square keep but these had not been successful, and in some cases produced a worse 'field of fire' than the square buildings they were meant to replace. In one other respect castles were taking a stride forward. Buttresses and vaulting were being introduced, neither of which had been employed in the castle-building of the previous century.

No account of Henry's reign is complete without some reference to the attractive and obliging Nesta of Carew Castle. Carew Castle, five miles north east from Pembroke, was owned by Gerald of Windsor, who was Henry I's castellan at Pembroke, after the expulsion of the Montgomery family. Carew may have been his wife Nesta's dowry. She was the daughter of the Prince of Deheubarth, and was very beautiful. She was taken to Henry's court as a hostage for her father, and obligingly bore Henry a son (Henry Fitzhenry). Gerald took her back and raised a family, but subsequently she was abducted by her cousin Owen Cadogan, Prince of Powys. Eventually the Prince of Powys was driven out of Wales, but a few years later returned and was killed by Gerald when they met by chance on a road. Gerald died in 1116 and Nesta, nothing daunted, married Stephen, the Constable of Cardigan, for whom she produced another family. Her families were Fitzhenries, Fitzgeralds, and Fitzstephens. All took part in the Conquest of Ireland and founded famous Irish dynasties. The Fitzgeralds in Wales eventually took to calling themselves Carew.

Henry I died of a surfeit of lampreys 'of which he was fond, though they always disagreed with him; and though his physician recommended him to abstain, the king would not submit to his salutary advice' (Henry of Huntingdon). Opinions of him were divided. Some said he had great sagacity, splendid achievements, and great wealth; others 'taking a different view attributed to him three gross vices, avarice . . . as he impoverished the people by taxes and exactions, cruelty, in that he plucked out the eyes of his kinsman the Earl of Morton in his captivity, and wantonness, for like Solomon he was perpetually enslaved by female seduction.'

His father had been lucky to be buried at all for his corpse had lain naked until it was buried as an act of charity. His brother, William Rufus, had had an undignified last journey on a charcoal burner's cart. But Henry was an English king and must be buried in England. Some time elapsed before the remains could be brought to Reading Abbey, and the mediaeval embalmer who tried to officiate was fatally infected in the process. The macabre details are recorded by contemporaries with some relish.

The lessons learnt on this occasion were remembered. When Henry V died in France nearly three hundred years later his body was boiled as a measure of preservation.

❋ 5 ❋

'The Nineteen Long Winters when God and His Saints slept'

Stephen and Matilda (1135–1154)

THE period which followed the harsh but stable rule of Henry I had the attributes of a nineteen-year-long nightmare. Henry left numerous descendants but the only surviving one born in wedlock was Matilda, now a widow but formerly married to the Emperor Henry V. The King's legitimate son, Prince William, had been drowned fifteen years before when a drunken skipper and crew ran the 'White Ship' on to rocks off the French coast. England took the death of the heir to the throne with stoicism, partly because the young Prince was known to be an unpleasing young tyrant, and partly because the three hundred passengers and crew from noblemen to deckhands were all believed to be homosexual, and drowning a very appropriate fate for them. However, as there were a number of women on board it may be that they were not included in the general condemnation.

Henry was fully determined that Matilda should succeed him although his illegitimate son, Robert of Gloucester, or his sister's son, Stephen of Blois, would have been more acceptable to the barons. The English had, in the past, accepted female rulers, but the Normans never. The fact that the leading barons had sworn to support Matilda after Henry's death meant nothing when the day came.

Unfortunately for Matilda, her cousin Stephen had obvious qualifications for the English crown, apart from the fact that his claim was virtually as good as her own. He was brave, attractive and popular, was married to a woman of Saxon descent, and already held vast estates in England. His most powerful asset lay in the fact that his brother Henry was Bishop of Winchester. The importance of this was shown when the Bishop not only obtained the keys to the Royal Treasury but also persuaded the two most influential men in the country to support Stephen. Accordingly Stephen was crowned before Henry I was buried. Unluckily for England, Stephen possessed a number of admirable qualities such as courage, generosity, and amiability, but lacked the one unpleasant quality necessary for his position; he had nothing of the ruthless egotism, tinged with avarice, that had marked his predecessors. In consequence the Norman barons were able to do much as they pleased on nicely calculated risks. For it is a hard fact about authority that the slight element of uncertainty, the unpredictable reaction, and the occasional touch of inhumanity, are extremely effective at keeping self-seekers out of mischief. Matilda, when her turn came, failed for precisely the opposite reasons. Her nature was harsh and mean, she had no attractive qualities, and she failed to understand that loyalty cannot be bred from hatred.

The mistake that rendered Stephen's position impossible was made very early in his reign, when it was rumoured that Robert of Gloucester, at this time in Normandy, was going to renounce his allegiance to Stephen. At this point he condoned the fortification of baronial strongholds, believing somewhat naively that the holders would resist Matilda's party and support him in consequence. The 'licence to crenellate', that is to fortify with battlements, could only be obtained from the Crown. Except for this short period of Stephen's reign it was never lightly given, and even then Stephen was partly tricked into the concession. William of Malmesbury puts it: 'For many people emboldened to illegal acts, either by nobility of descent or ambition, or rather by the unbridled heat of youth, were not ashamed to demand castles, other estates, and indeed whatever came into their fancy from the King. When he delayed complying with their requests . . . they, becoming enraged immediately, fortified their castles against him, and drove away large booties from his lands. Nor indeed was his spirit at all broken by the revolt of any, but attacking them suddenly in different places he always concluded matters more to his own disadvantage than theirs; for after many great and fruitless labours, he gained from them, by the grant of honours or castles, a peace feigned only for a time.'

With a weak king on the throne, a strongly supported claimant across the channel, and a host of ambitious local tyrants, it was inevitable that the weak would suffer. They had not long to wait.

Two entirely private rebellions broke out in 1136. First Hugh Bigod, Earl of Norfolk, calmly captured the royal castle of Norwich. Then the city of Exeter was pillaged on the sole excuse that King Stephen was rumoured to be dead—a story that emanated from Baldwin de Rivers (or Redvers). When William I had conquered Exeter in 1068 he had commissioned one Baldwin de Brionne, Sheriff of Devonshire, to build a castle capable of overawing the city. Baldwin de Brionne had taken the northern quarter of the city, which sloped steeply to the north and west, and built a substantial motte and bailey wooden castle. Soon the first construction was replaced by a large shell keep and a powerful gatehouse with walls six feet thick. By Stephen's reign the castle had also acquired high towers. But the fact that Baldwin de Brionne had built it on royal command did not influence his grandson Baldwin de Rivers in the slightest; on the contrary he held it stubbornly for three months against the Conqueror's grandson.

The story of the siege is vividly described in *Gesta Stephani*, an anonymous work by a contemporary of King Stephen. He describes the castle as held by the 'flower of all England', all of whom had taken an oath to resist the King. The citizens of Exeter were so far unconvinced that Stephen was dead that they had sent messages to him to relieve them of Baldwin's plundering. The King arrived with an impressive retinue and was received with joy by the citizens and derision from the castle. Aggressive defiance included well-aimed javelins, and unexpected sallies from hidden posterns; the latter were somewhat of a novelty in Britain and show how advanced in design Exeter Castle was. Stephen pressed home the attack with great vigour. 'For with a body of foot soldiers, very completely equipped, he resolutely drove the enemy back and took an outwork raised on a very high mound to defend the castle, and he manfully broke an inner bridge. . . . Sometimes he joined battle with them by means of armed men crawling up the mound; sometimes by the aid of countless slingers, who had been hired from a distant region, he assailed them with an unendurable hail of stones; at other times he summoned those who have skill in mining underground, and ordered them to search into the bowels of the earth with a view to demolishing the walls; frequently too he devised engines of different sort, some rising high in air, others low on the ground, the former to spy out what was going on in the castle, the latter to shake or undermine the wall' (*Gesta Stephani*).

The 'countless slingers' were probably mercenaries from the Balearic islands, where the sling was said to have been invented. According to Vegetius in *De Re Militari* Balearic mothers did not allow their children to have their food until they had hit it with a stone from a sling. Presumably they were weaned early. Slingers were apparently feared more than archers at this time. They never ran short of ammunition; a blow from a slung stone was often fatal; the lightness of the weapon enabled them to be the most mobile of attackers; and their fire seemed as fast as hailstones.

However, slingers did not bring Stephen victory at Exeter. After three months the siege had cost the King 15,000 marks (less than fifty years before the Duchy of Normandy had been mortgaged for 10,000) and such sums were not easily forthcoming. At this point both the castle wells dried up, a fact attributed to the direct intervention of Providence, for they had never failed before even in the most severe droughts. Wine was then used as a substitute, for drinking, for making bread, and for extinguishing fires lighted by incendiary arrows. The latter practice soon consumed the wine stocks. Agonies of thirst then caused the garrison to seek peace terms, but their extreme physical discomfort did not blunt their cunning. Two emissaries were sent 'first in rank and dignity, men skilled to adorn their speech with charm, and give their words, whenever it suited them, the turn that wisdom and elegance most required'. However their charm did not daze the sharp eye of the Bishop of Winchester, who stood at Stephen's elbow and soon guessed from their looks that they were parched with thirst. They were sent away to fight on. Baldwin's wife then followed, barefoot and in tears, but to no avail. Eventually Stephen's own supporters prevailed on him to let them surrender; unfortunately for the King they persuaded him that the rebels should not be punished. As Henry of Huntingdon puts it, 'being ill advised he permitted the rebels to go without punishment, whereas if he had inflicted it so many castles would not have been held against him'. Grim though the latter stages of this siege were, they were slight compared with the siege of Xerigordo described by the anonymous author of *Gesta Francorum*.

Following his hollow success at Exeter Stephen promptly created fresh trouble for himself at Bedford. The castle was held by one Miles de Beauchamp who bitterly resented it being given to the man who had married his cousin. *Gesta Stephani* describes it as being 'surrounded by a very lofty mound, encircled by a strong and high wall, fortified with a strong and unshakable keep, and filled with tough and unconquerable men'. Initially it was attacked with a variety of engines, carefully watched, kept under an almost continuous hail of arrows, and generally harassed; all without the slightest effect on its resistance. Subsequently Stephen was called away to attend to other urgent matters, and left a passive force blockading Miles in his citadel. Exhaustion soon combined with shortage of food to bring Bedford Castle to surrender.

From Bedford Stephen had to rush back to the west country. Bristol was the stronghold of Matilda's supporters and was an exceptional problem for the besieger. The main difficulty lay in the fact that it could not properly be surrounded. Suggestions such as blocking the mouth of the harbour with rocks and turf were dismissed, quite rightly, as being impracticable, and Stephen broke off the siege. Instead he tackled Castle Cary and Harptree; the former he took by starvation, the latter by the stratagem of appearing to abandon the siege. Somewhat stupidly the garrison at Harptree assumed when he retired that he was going to besiege Bristol; they rushed after him, were completely fooled when he doubled on his tracks, and apparently lacked the initiative to take him in the rear when he once more assailed the castle and took it.

The stalemate principle was introduced when Stephen moved to crack a very tough nut in the castle of Dunster. A glance was sufficient to tell him that his task here was as hopeless as at Bristol. Instead of wasting time and energy in a siege, he merely built another castle, presumably of wood, in front of Dunster, and left it in charge of Henry de Tracy, a man well-qualified for the task. De Tracy thereupon established himself as lord of the district, harassed the garrison of Dunster, and effectively prevented it from raiding the surrounding area. As its castellan William de Mohun had previously earned the title the 'scourge of the west', De Tracy's feat was not inconsiderable.

By 1138 Robert of Gloucester had organized a baronial rebellion in the south to coincide with an attack by Matilda's uncle, King David of Scotland. Of the people who caused endless misery to the English people at this time, these were perhaps the only disinterested ones.

The Scottish invasion was launched in March 1138. The Scots army was so wild and savage that the northern barons were temporarily shocked into forgetting some of their own ambitions. Stephen was too far away to be of immediate use and the defence of the north had to be organized by others. The incredible character who emerged at this point was Thurston, Archbishop of York. He was old, half-crippled, and ill, but he organized an army with speed and efficiency that few in history have equalled. He would have taken part in the battle himself if he had been able to get into his armour, but sickness and weakness prevented this last demonstration of leadership quality. The vital battle at Northallerton, known as the Battle of the Standard, stopped the Scots, who were reputed to have lost 10,000 but it did not finish the war; hostilities continued for another year and were only concluded by the cession to the Scots of Northumberland, Cumberland, and Westmorland. Henry of Huntingdon gives the figure at 11,000 casualties on the battlefield, and a wholesale massacre of fugitives afterwards.

While the north was doing as best it could with the Scots, Stephen had his hands full in the west. His troubles were largely of his own making. When Matilda had first landed in Sussex he had swiftly surrounded and captured her in Arundel Castle. Then, with chivalry gone berserk he allowed her to join her half-brother Robert of Gloucester at Bristol, to which she was escorted by Stephen's brother, Henry of Winchester.

The author of *Gesta Stephani*, a great admirer of Stephen, justified this move by saying that it concentrated the opposition in one area and made it easier to conquer. Events soon disproved this theory. The first move appears to have come from Wallingford where the holder, one Brian Fitzcount, 'a man of distinguished birth and splendid position . . . rebelled against the King, assisted by a large body of soldiers'. While Stephen besieged Wallingford a host of others imitated Fitzcount's example. To make matters worse Stephen's army was so harassed by sorties from Wallingford Castle, which he had no hope of taking, that the King had to move elsewhere for his own safety. Before leaving he erected two castles to act as malvoisins and check the marauding activities of the Wallingford garrison. He then set off to capture Trowbridge, and had some luck on the way by capturing Cerney and Malmesbury without difficulty. But while he was heading for Trowbridge its rebel castellan Miles marched swiftly and neatly round Stephen's army and attacked the newly erected castles at Wallingford. It appears to have been a night attack and resulted in total victory. From this point Miles went from strength to strength and soon dominated the west. Stephen was unable to capture Trowbridge and contented himself with leaving a garrison at Devizes, 'a body of troops very prompt in battle, and as the two sides assaulted each other with alternate raids they reduced all the surrounding country everywhere to a lamentable desert'.

In 1139 he besieged and took Leeds Castle which was not, as might be supposed, in Yorkshire, but in Kent between Maidstone and Ashford. Subsequently he campaigned in the north-west and Scotland, from the latter of which he removed Henry, the son of King David. This young man accompanied Stephen on his campaigns, and had some interesting adventures. First, at the siege of Ludlow he was nearly captured by an iron hook let down swiftly from the walls; he was saved from this undignified fate by Stephen's strength, swiftness, and presence of mind. The hook was presumably intended to catch Stephen. (Plate 4 gives an idea of the device used by the defence on this occasion.) On returning from the siege he met Ada, daughter of William de Warenne, Earl of Surrey, and soon after married her. They had six children, two of whom, Malcolm and William, became Kings of Scotland. At about this time Stephen had to besiege Shrewsbury Castle. Although the castellan and garrison were defeated their manners were so arrogant that Stephen lost his temper and hanged 93 of them, including the Constable. The effect was electric, and for a while subservience greeted Stephen everywhere: had he kept up this pitch of severity it was obvious that his troubles would soon have disappeared.

Unfortunately he chose this moment to alienate the Bishops, who, in the best Norman tradition, were more adept with the sword than the word. He had little reason to trust them and suspected that their recent activities in castle-building and strengthening were preparations for welcoming Matilda. But he went too far for his supporters, even for his own brother, Henry, Bishop of Winchester. First Stephen imprisoned the Bishop of Salisbury, a man of great wealth and power but little piety, then the Bishop of Lincoln who was believed to be the son of the celibate Bishop of Salisbury. Both died soon afterwards and their goods were redistributed; although some doubtless found their way into the royal coffers to pay for this exhausting series of wars.

The Bishop of Ely, nephew to Salisbury, was no less worldly than his uncle, but younger and more active. Enraged by Stephen's act, and fearing similar treatment 'he put on the man of blood and after hiring in Ely at his own expense knights who were prepared for any crime, ready in hand and mind, he molested all his neighbours and especially those who supported the King. Now Ely is an agreeable island, large and thickly inhabited . . . impenetrably surrounded on all sides by meres and fens, accessible only in one place, where a very narrow track affords the scantiest of entries to the island, and the castle, wondrously set long since, right in the water, in the middle of the opening of the track, makes one impregnable castle of the whole island.'

'The King . . . hastily arrived here with a large army, and after examining the wonderful and unconquerable foundations of the place he anxiously consulted a number of persons about the means of breaking in with his men.'

He would have been saved much time and trouble if he had learnt about the campaign of his grandfather William I at the same place; his own followed much the same lines through trial and error. But if Stephen was no Conqueror his opponent was no Hereward. On arriving at Ely with a large army Stephen decided the island was virtually impregnable so spent some time in consultation before committing his forces. Eventually he accepted the recommendation that he should collect a number of boats and build a pontoon bridge on them with hurdles. This manoeuvre took him to the island but not to the interior. However, a monk came forward to act as a guide (thereby betraying his bishop) and led Stephen into the interior; the Bishop of Ely thereupon fled. The treacherous monk was rewarded by being made an abbot but apparently endured 'many toils and afflictions through God's just judgement of what he did in secret'.

For a time matters went well with Stephen. He reduced large areas of the west country and was apparently establishing a measure of stability, when he lost Devizes Castle. Devizes was thought to be impregnable, but fell to a night attack effected entirely by scaling-ladders. Surprise was complete, and so was victory. It was a remarkable feat accomplished by one Robert Fitzhubert, a Fleming. However, Fitzhubert was so inflated by self-importance after the capture that he refused to surrender Devizes to his overlord, the Earl of Gloucester. The latter had a loyal and cunning supporter who held the castle at Marlborough to which Fitzhubert was decoyed and imprisoned; soon after he was hanged in front of Devizes Castle, which had been the scene of his recent triumph.

The full effects of these events were not seen until 1141, when the Earl of Chester decided to change sides and capture Lincoln. The manner in which this was accomplished was unusual. 'Cautiously choosing a time when the garrison of the tower were dispersed abroad and engaged in sports they sent their wives before them to the castle under pretence of their taking some amusement. While, however, the two countesses stayed there talking and joking with the wife of the knight whose duty it was to defend the tower, the Earl of Chester came in, without his armour or even his mantle, apparently to fetch back his wife, attended by three soldiers, no one suspecting any fraud. Having thus gained an entrance they quickly laid hold of the bars, and such weapons as were at hand, and forcibly ejected the King's guard. They then let in Earl William and his men at arms, as it had been planned before, and in that way the two brothers got possession of the tower and the whole city.'

The people of Lincoln remained loyal to Stephen and sent him news of what had occurred. After a swift march he was able, with local help, to put the castle under siege, but his troops were not sufficiently vigilant, and the Earl of Chester slipped through the besiegers by night, and hastened to Robert of Gloucester. The latter assembled a large army, which included a substantial Welsh contingent with more courage than weapons. As this moved nearer to Lincoln, Stephen vacillated between various plans, not knowing whether to carry on with the siege, to defend the town, or give open battle. Ultimately he decided on the latter but his dispositions were of little avail as large numbers of his troops deserted to the enemy. Before this battle, as in others, every attempt was made to convince the combatants of the rightness of their cause, the certainty of victory, and the worthlessness of the opposition. Henry of Huntingdon gives a full account of these propaganda speeches. Robert of Gloucester described the King's supporters as 'Alan of Brittany, a man so execrable, so polluted with every sort of wickedness, who never lost an opportunity of doing evil, and who would think it his deepest disgrace if anyone else could be put in comparison with him for cruelty, the Earl of Mellent, crafty, perfidious, whose heart is naturally imbued with dishonesty, his tongue with fraud, his bearing with cowardice, the Earl of Albemarle . . . a man from whom his wife was compelled to become a fugitive on account of his intolerable filthiness. The earl also marches against us who carried off the countess just named; a most flagrant adulterer, and a most eminent bawd, a slave to Bacchus, but no friend to Mars, redolent of wine, indolent in war . . .' And so on.

As Stephen's voice was neither clear nor strong Baldwin Fitzgilbert replied for him. Gloucester was credited with the mouth of a lion but the heart of a hare, Chester with the qualities of an idiot, and the remainder not much better. The battle which followed was swift and devastating. In spite of desperate personal efforts Stephen was captured and taken to a dungeon in Bristol Castle. William of Malmesbury says he was treated courteously but others report that he was loaded with chains. In any event he was more fortunate than the wretched inhabitants of Lincoln who were 'butchered like cattle and put to death in various ways without the slightest pity'. Apart from the subsequent brutality there were two other features of this encounter which showed how high feelings were now running. So desperate had Earl Robert been to get at Stephen that he and his army, quoted as ten thousand, had swum the unfordable, flooded, river Trent; equally Stephen's supporters had turned the encounter into an English infantry battle by fighting dismounted and hand-to-hand.

At this point it seemed as if nothing could now prevent Matilda's complete triumph; but by a combination of greed and haughty stupidity she contrived to throw away the opportunity of her lifetime.

As Stephen had already been crowned there had to be some preparation before Matilda could take the same position. The main problem was the Church but this was resolved by skilful manoeuvring on the part of Henry of Winchester, who had no compunction about abandoning his brother's cause. A lesser problem was the attitude of the Londoners but Robert of Gloucester made diplomatic preparations in that area and it seemed that all would be well.

Such optimism reckoned without Matilda. On entering the city she promptly imposed a heavy fine on the citizens as a penalty for their support of Stephen, and followed it up with acts calculated to anger her friends as well as her enemies. The climax to this came on an afternoon of astonishing activity. The citizens, apparently under the impression that Stephen had returned, sprung to arms in a spontaneous riot. Matilda, equally surprised, rushed from the dinner-table, leapt on a horse, and galloped to safety with nothing but the clothes she was wearing.

With London lost, Matilda busied herself with improving her position from Oxford. Before long she had reasonable doubts about the fidelity of Bishop Henry and set off to Winchester to resolve them. The wily Bishop moved out of the town as Matilda moved in, allowed her sufficient time to occupy the royal castle, and then returned and besieged it. As it also held Robert of Gloucester and the King of Scotland, as well as their chief supporters, this was nemesis indeed.

'The Bishop, sending all over England for the barons who had obeyed the King, and also hiring ordinary knights at very great expense, devoted all his efforts to harassing them outside

THE MAIN AREAS OF THE CONFLICT BETWEEN
STEPHEN AND MATILDA

Figure 17.

the town. The queen likewise with a splendid body of troops, magnificently equipped, besieged the inner ring of besiegers from outside with the greatest energy and spirit'. As *Gesta Stephani* says, 'It was a remarkable siege, nothing like it was ever heard of in our times. The whole of England, together with an extraordinary number of foreigners, had assembled from every quarter, and was there in arms, and the roles of the combatants were reversed in so far as the inner besiegers of the bishop's castle were themselves very closely besieged on the outside by the king's forces, never without danger to men, never without the heaviest loss to both parties. To say nothing of the knights on the one side or the other who were being taken in the daily fighting or were drawn by different fates to meet different deaths.'

Most of the town, including two abbeys, was burnt to ashes, and famine prevailed among the surviving citizens. After six weeks of siege a desperate break-out was effected in the early hours, and Matilda reached Devizes castle. Gloucester, not so lucky, was captured, and taken off to Rochester. Matilda, whatever her faults diplomatically, was an exceptional woman, and in order to get to the safer castle at Bristol got into a coffin and was carried there on a bier as an alleged corpse. There is an alternative story that after a forty mile ride she had to be carried like a corpse on a litter between two horses.

By the end of the year Stephen had been bartered for Robert of Gloucester, and the original miserable conflict was resumed.

After the disasters of Winchester, Matilda took a little time to recover but she soon took up a strong defensive position at Oxford. This in itself was formidable, 'inaccessible because of the very deep water that washes it all round, most carefully encircled by the palisade of an outwork on one side, and on another finely and very strongly fortified by an impregnable castle and a tower of great height' (*Gesta Stephani*). From Oxford she arranged a defensive screen through castles at Radcot (so surrounded by water and marsh as to be inaccessible), Woodstock, Cirencester, and Bampton. Stephen captured or bypassed these, and with considerable personal gallantry (he swam the river under fire) led the charge into the city. Matilda was accordingly besieged in the castle which Stephen made every effort to reduce. Siege engines battered the walls continuously, pickets watched every approach road by night and by day, and within three months the garrison was on the verge of starvation.

At this point Matilda, resourceful as ever, chose a night when the country was blanketed with snow, and the rivers frozen, dressed herself in white, and escaped with three chosen knights. Some chroniclers say she let herself out through a postern but the best documented story is that she was let down by a rope from the keep, found her way through pickets, and walked over the ice and snow six miles down the river to Abingdon. At Abingdon she was able to obtain a horse and rode a further ten miles to sanctuary at Wallingford. At Wallingford was a large army. In the *Historia Novella* William of Malmesbury says 'the nobles of the Empress's party, ashamed at having left their lady contrary to the agreement, massed their forces at Wallingford with the intention of attacking the King if he would risk a fight in the open field, but it was not their plan to assail him within the city, which the Earl of

Gloucester had so strongly fortified with earthworks that it seemed impregnable unless it were set on fire.'

Very little trace of these earthworks now remains.

In 1143, two years after his frustrating siege of Oxford, Stephen once more besieged Lincoln. He did not make much progress but the event shows the extent to which mining was used. According to Henry of Huntingdon, 'while he was preparing a work for the attack of the castle, eighty of his workmen were suffocated in the trenches, whereupon the King broke up the siege in confusion'. In the same year his great enemy Miles of Gloucester died from an arrow wound. The choicest fate was reserved for one Robert Fitz Hildebrand 'a man of low birth, but also of tried military qualities . . . likewise a lustful man, drunken and unchaste'. When William Pont de l'Arche quarrelled with the Bishop of Winchester and changed sides to Matilda's, Fitz Hildebrand was sent with troops to reinforce his garrison. He became very friendly with William but even more so with his wife. 'Stung with desire, he seduced his wife, and afterwards when a vile and abominable plan had been formed by agreement between him and the wife, he fettered William very tightly and imprisoned him in a dungeon, and enjoying his castle, wealth, and wife he likewise abandoned and rejected the countess who had sent him there, and made a pact with the King and the bishop. Nor did that reckless seducer escape punishment . . . a worm was born at the time when the traitorous corrupter lay in the unchaste bosom of the adulteress and crept through his vitals, and slowly eating away his entrails it gradually consumed the scoundrel' (*Gesta Stephani*).

Suitable afflictions also rewarded other unseemly activities. Robert Marmion castellated Coventry Abbey and when defending it sallied forth surrounded by troops; he was the only one to be killed. Godfrey de Mandeville seized and fortified Ramsey Abbey; appropriately he was shot through the foot by a common soldier. Godfrey's son was thrown from his horse and died from concussion. Reiner, his chief officer, was drowned; the ship he was travelling on stuck fast in the middle of the sea. Lots were cast to discover the cause. Three times they singled out Reiner, who was thereupon put into a small boat. The small boat promptly sank, but the ship, now freed of its evil burden, came unstuck and sailed on happily (it was said).

But these were isolated instances. For every one criminal who disappeared there were ten to replace him. The country swarmed with foreign mercenaries 'who were unceasingly occupied in pillaging the goods of the poor . . . encouraging hostility on both sides, and murdering men in every quarter'. The Bishops who should have been checking such activities, indeed, took part in them, and it was said that the Bishops of Winchester, Lincoln, and Chester were the worst of all. An exception was the Bishop of Hereford who defied and excommunicated Miles, formerly of Gloucester, now Earl of Hereford. The latter was completely overawed by the Bishop's vigour, and agreed to restore all the church property he had acquired. Shortly afterwards Miles was killed in a hunting accident, being shot through the chest in much the same way as William Rufus. Another archvillain to be despatched by an arrow was Geoffrey de Mandeville who had unwisely taken off his helmet to mop his brow. Like many arrow wounds it did not appear to be serious at the time but soon proved fatal.

In 1142 one William de Dover, 'a man crafty and bold in warfare, relying on the support of the Earl of Gloucester, came to the village named Cricklade, which is situated in a delightful spot abounding in resources of every kind, and built a castle which was inaccessible because of the barrier of water and marsh on every side. . . .' Cricklade is on the upper Thames, and is now well-drained but in those days was probably nearly as inaccessible as Ely. De Dover proceeded to harass the neighbouring royal castles with such vigour that Stephen felt obliged to take action. However, Robert of Gloucester amassed such a 'cruel and savage army of footmen from Wales and Bristol' that Stephen was outnumbered and withdrew. On the way back he took Winchcomb in a desperate onslaught, the orders being that 'some should advance shooting clouds of arrows, others should crawl up the mound, and everyone else should rush rapidly round the fortifications and throw in anything that came to hand'. Shortly afterwards De Dover repented, 'went to the holy places of Jerusalem to expiate his sins and there, after manfully doing many glorious deeds against the persistent enemies of the Christian faith, at last he was killed and died a blessed death' (*Gesta Stephani*).

Robert of Gloucester then built a castle at Farringdon (of

which no trace now remains), which in 1146 provided a vigorous and sophisticated siege. First Stephen surrounded the castle, then instructed his army to build a stockade protecting them from sudden sallies. And 'setting up engines most skilfully contrived around the castle, and posting an encircling ring of archers in dense formation, he began to harass the besieged most grievously. On the one hand stones and other missiles launched from the engines were falling and battering them everywhere, on the other a most fearful hail of arrows flying around before their eyes, was causing them extreme affliction; sometimes javelins flung from a distance, or masses of any sort hurled in by hand, were tormenting them, sometimes sturdy warriors gallantly climbing the steep and lofty rampart, met them in most bitter conflict with nothing but the palisade to keep the two sides apart.'

This was Stephen's greatest moment. With Farringdon he won not only a castle and considerable arms and treasure, but also the respect of his bitterest enemies. Randulf, Earl of Chester long a thorn in his flesh, now joined and helped him. He captured and handed over the town of Bedford, and then joined forces with the King against Wallingford. Together they built 'within sight of Wallingford a castle that was a work of wondrous toil and skill'; the aim of this was to check the activities of the Wallingford garrison. However, Chester soon tired of being a friend and reverted to being an enemy—a part which he found more congenial. He concocted an ingenious plan to decoy Stephen's army into a remote part of Wales where it could be easily massacred, but the plot misfired and Chester was imprisoned at Northampton. With amazing rashness Stephen accepted sureties of good behaviour and freed him; needless to say, Chester then embarked on a series of offensives that made his previous career seem tranquil. He lived till 1153 although he had a narrow escape when an attempt was made to poison him; his three attendants died, but he survived because he had not drunk as much as they had.

In 1147 Matilda's 16-year-old son landed in England. Most rashly he did not arrive with a large army, but with a small party who had been hired on promises. It was a discouraging start for the future Henry II.

When he tried to attack Cricklade he was put to flight.

When he assaulted Bourton his army dispersed in panic. Desperate for money he appealed to his mother, but she was in the same straits; his uncle, Robert of Gloucester, had money but preferred to retain it. And then—the supreme example of the extraordinary attitudes of the age—he appealed to his cousin King Stephen to lend him money to fight against him. Stephen, of course, sent him some, although his own son Eustace, a most promising young man, was still alive at this time. Gloucester, implacably hostile to Stephen to the end, died of fever in 1147, leaving as his successor 'William, already advanced in years but effeminate and more devoted to amorous intrigue than war'. Matilda, with her main support gone, lost heart and retired to Anjou.

The end was not quite in sight. Castles were built, besieged, and taken, crops were burnt, innocent people were slaughtered. Stephen marched about the country putting down insurrections only to find another had broken out behind him or in a distant corner.

In 1152 Matilda's husband, the Count of Anjou, died: and young Henry succeeded him as Count of Anjou and Duke of Normandy. As such he was a formidable military proposition. He then married the enormously wealthy Eleanor of Aquitaine, divorced wife of the King of France; her superb abilities as a trouble maker were so far undeveloped. Almost immediately Henry was involved in a war with the King of France, who allied himself with his son-in-law, Eustace, Stephen's son. This alliance placed Stephen in a slightly better position so he decided to tackle Wallingford again, for long an obstacle to his plans.

First he built two castles, secondly he seized the bridge that now supplemented the ford. An attempted sally was beaten back with heavy casualties. Henry of Huntingdon says that Stephen built a fort on the bridge at the entrance, which prevented all ingress, so that provisions could not be introduced. Beginning to feel the pressure the garrison petitioned their lord (i.e. Henry) that he would either send them relief or they might have licence to surrender the castle into the King's hands.

Henry decided to return, and after a difficult sea voyage marched to Malmesbury and attacked it. He took all of the castle but the keep, which proved impregnable. Stephen tried to relieve this gallant last bastion but when he met Henry's army the potential battlefield was drowned with torrents of rain. As it drove into the faces of Stephen's troops, and as the river was too flooded for him to ford it, he abandoned the attempt and retreated to London.

The keep at Malmesbury thereupon surrendered and Henry marched adroitly to the relief of Wallingford. He went first to Crowmarsh which is on the opposite side of the river, and where there are traces of earthworks on the bank today. He laid siege to the castle there 'commencing the difficult and important enterprise by digging a deep trench round the walls and his own camp, so that his army had no egress but by the castle at Wallingford, and the besieged had none whatever'. However, his first attempt at attack was repulsed ignominously. It seems that the defence had concealed itself in the neighbourhood of the castle and when Henry began his assault, emerged and besieged him. Having extricated himself from this humiliating position he collected an even larger force: and Stephen in his turn called up every man he could. A crisis had been reached in the struggle transforming its character. Now it was not so much a siege as a battle array, with a river separating the armies. Henry even went so far as to level the earthworks he had just raised, presumably with the object of clearing the field of fire. At this point the barons on both sides decided that the whole affair had become too dangerous. This was not going to be a siege which could be broken off at will but a desperate conflict which would exterminate most of the nobility. Accordingly they persuaded Henry and Stephen to parley in secret. This extraordinary meeting took place by a little rivulet but they could agree on nothing, and perhaps discussed nothing but the faithlessness of the barons, who preferred anarchy to authority. Having retired without any agreement, both began further sieges elsewhere.

Then Stephen's son Eustace died suddenly of a fever. Here was a chance for a diplomatic settlement, and in November, 1153, the Bishop of Winchester arranged for Stephen to have the crown in his lifetime on the condition that Henry was named his heir. Although now the accepted King, Stephen still had to busy himself with sieges. Within a year he too had died of a fever, caught at Dover, and anarchy was at an end.

In Stephen's reign chaos had become a permanent state. It is doubtful whether conditions were uniformly bad all over the country but the descriptions given in the Anglo-Saxon Chronicle paint a grim picture of feudal anarchy at its worst. 'Every powerful man built his castles and held them against him [Stephen] and they filled the country with castles. They oppressed the wretched people of the country severely with castle building. When the castles were built, they filled them with devils and wicked men. Then both by night and day they took those people they thought had any goods—men and women, and put them in prison and tortured them with indescribable torture to extort gold and silver—for no martyrs were ever tortured as they were. They were hung by the thumbs or by the head, and corselets were hung on their feet. Knotted ropes were put round their heads and twisted till they penetrated to the brains. They put them in prisons where there were adders and snakes and toads, and killed them like that. Some they put in a "torture chamber", that is a chest that was short, narrow and shallow, and they put sharp stones in it and pressed a man so that he had all his limbs broken. In many of the castles was a "noose and trap"—consisting of chains of such a kind that two or three men had enough to do to carry one. It was so made that it was fastened to a beam, and they used to put a sharp iron around the man's throat and neck, so that he could not in any direction sit or sleep but had to carry all that iron. Many thousands they killed by starvation.'

Among the other activities of the time 'they levied taxes on the village every so often, and called it "protection money"'. J. H. Round in *Geoffrey de Mandeville* quotes *Monasticon* IV 142 and other sources of how barons' agents, disguised as beggars, wormed out secrets of scanty hoards; the owners were subsequently tortured until they gave everything.

Of the last year of Stephen's reign the best that can be said is that conditions grew no worse; but this may well be because they could not. There were one or two brushes with the King-elect because Stephen indulgently spared some of the illegal castles. However, it should be noticed that, in spite of anarchy, the teaching of law began at Oxford in 1149, and numerous religious houses were founded in Stephen's reign.

Stephen was the last of the Norman kings. Henry was an

Angevin, the first of the Plantagenet line which took their name from the 'Planta genista' or sprig of broom they wore as their crest. There is no exact modern equivalent for the crest for it was more personal than insignia of rank.

Having experienced the sour ferocity of William I, the debauchery of William II, the chilly severity of Henry I, and the reckless stupidity of Stephen, the people of England looked forward to better things and were not at first disappointed.

❋ 6 ❋

The Plantagenet Warriors

Henry II (1154–1189): Richard I (1189–1199)

HENRY II did not land in England until six weeks after his succession to the throne. The news of Stephen's death found him engaged in the siege of Torigny and, when this was concluded, he still had to settle some local disturbances before he was free to leave Normandy. Even after these continental affairs had been decided, his sailing was further delayed by storms and adverse winds.

The tasks that faced this hot-tempered but cool-headed young man might well have daunted anyone. But Henry, although only 21, was not without experience, and he knew that he had considerable power behind him. Not unnaturally, he saw himself as a French prince with an English kingdom. Apart from Anjou and Normandy he also had the vast territory of Aquitaine, the inheritance of his wife Eleanor, thirteen years his senior and described by contemporaries as a harlot. Subsequently he would have cause to regret having married this erratic lady but in the initial stages of his reign her dowry was an enormous advantage.

His English inheritance was not a kingdom but a state of chaos. Law and administration had disappeared: there was no money and no immediate possibility of producing any: and the country was dotted with illegal fortifications. However, the military side of his activities occupied less time than the other reforms which were necessarily bound to be slower.

First he ordered the dismantling of all the illegal castles that had been built during the previous nineteen years. Some of them were occupied by foreign mercenaries who had been brought over by Stephen. In expelling these he was careful to make the act legal, not a mere military demonstration; they were well-hated and both English and Normans were glad to see them go.

Reducing the network of castles was a formidable task so he raised a large army for the purpose. Known as 'adulterine' castles, these buildings varied greatly in strength. Some were solid stone buildings, others were motte and bailey structures, others were hardly castles at all. He was said to have destroyed over eleven hundred adulterine castles but it is unlikely that more than half of these were recognizable in the known sense of the word. Many offered no resistance, others were easily starved into surrender, but three, Bridgenorth, Cleobury, and Mortimer had to be taken by vigorous attack. Like his son Richard, he was reckless of personal safety, and nearly met a similar fate. In 1155, when besieging Bridgenorth, he had a narrow escape from a carefully-aimed arrow. Fortunately for Henry it was intercepted by a knight of his bodyguard, Hubert de Clair, who died of the wound. In a campaign taking approximately a year, he imposed royal control from one end of the country to the other, and finally compelled the King of Scotland to give up the counties of Northumberland, Westmorland, and Cumberland, ceded to him in the reign before. For good measure, the six castles of Henry, Bishop of Winchester, all powerful structures, were flattened to the ground, while their owner fled to refuge at Cluny.

Following the completion of his campaign in England he was about to tackle Ireland when he was recalled to France. The cause was treachery. His younger brother had decided that as Henry had become King of England he should relinquish to the next-born the title of Count of Anjou and Maine and had taken possession. A brisk campaign settled the issue and Henry magnanimously granted a generous pension to his discontented brother.

On his return to England in 1157 the young king set about establishing his rule in Wales. In the early stages of the campaign he was surprised to find no opposition but he was even more surprised a little later when he was attacked in a narrow defile in Cynsyllt. A similar situation would occur in Pilleth in 1402, but on the later occasion defeat would be absolute. Henry II, on the other hand, was able to fight his way out, although with crippling losses. Subsequently he kept close to the coast, exercising caution, and erecting castles along his route. Ultimately the Welsh came to terms, swore homage, and gave up land illegally taken in Stephen's time.

In 1157 he undertook an expedition to France to gain Toulouse, on the strength of a very dubious claim. The expedition was comparatively fruitless, for its diplomatic disadvantages caused him to abandon it, but it led to a change in military service that would have far-reaching effects.

Until 1157 Norman warfare had been conducted by forces contributed under the feudal obligation of forty days military service, supplemented by a few itinerant mercenaries. The forty-day service sufficed for minor operations but was liable to cause disruption and disaffection when a campaign dragged on or a castle refused to surrender. In fact, if a castle could hold out for six weeks it could well find the besieging force was thinned out sufficiently for an attempt at sortie. Henry, realizing that time alone would prevent him using forty-day service for wars in distant Aquitaine, introduced 'scutage', by which a knight's service could be commuted for a sum of money. By this means it would be possible to employ an army of mercenaries who would already be trained and equipped, who would not wish to go home half way through the campaign, and who would be professionals in every sense of the word.

Historically much of the interest in Henry's reign centres around the struggle between church and king. This was connected with the development of justice because Henry had decided, by the Constitutions of Clarendon, to make secular law supreme. In doing this he became involved in the long dispute with Thomas à Becket but was less hampered by the Archbishop himself than by the fact that the Pope was in the background. From a military point of view alone it would be folly to alienate the Pope, particularly for a king with possessions on the Continent. In the circumstances the Pope might place an entire country under an Interdict, which carried two potential dangers: legalized rebellion and resentment of his people against the offending monarch. This was later to occur in John's reign.

The murder of the Archbishop was undoubtedly a shock to Henry and as part of his expiation he undertook a form of crusade in Ireland. One result was to put the Irish under Papal governance, another was to put them under English rule. Whatever the rights of this conquest, nothing, in fact, could have been worse than their previous state. Five central kingdoms were surrounded by a number of smaller ones and vicious, senseless tribal wars occurred constantly. In spite of this semi-barbarous life the Irish had certain qualities that might in a later age have been called sporting or even civilized. They loved freedom and loathed work; leisure was much admired, and was not tainted with the slurs implicit in words like 'lazy' and 'idle'. They were faithful, if erratic, friends, and brave, dangerous enemies. They fought at close quarters with a steel hatchet called a 'sparthe', a sword called a 'skene', and two short javelins. They thought armour at worst a coward's trick, and at best a mistaken idea; this they often proved by getting to close-quarters and hacking through the mailed joints with the sparthe.

Henry's intervention was preceded by local dissension of more vigour than usual. Dermot, King of Leinster, had abducted the wife of a rival named O'Rourke but had subsequently been driven out of the country. With his affairs at this low ebb he had appealed in person for help from Henry. The latter had his own troubles at the time but allowed Dermot to recruit help from the English barony. Dermot enlisted Strongbow, Earl of Pembroke, and two remarkable Welshmen named Fitzstephen and Fitzgerald. They were sons by different fathers of the versatile Welsh Princess called Nesta mentioned earlier, who had once been mistress of Henry I, bearing to him Robert, Earl of Gloucester, the supporter of Matilda.

With 140 knights and 300 archers this force landed in Ireland, and was joined by a number of Dermot's supporters. By drawing the opposition out of the marshes and woods Strongbow's forces achieved a tremendous victory that culminated in a pile of 200 heads being made in front of Dermot who danced with joy at the sight. Subsequently Strongbow's forces went from victory to victory. As Dermot died suddenly at the height of the campaign Strongbow found himself Lord of Ireland. The latter part of Strongbow's campaign was concerned with Dublin. Dermot had stormed the city before he died. Most of the Dubliners had escaped when it was taken but they soon returned, aided by sixty Norwegian ships. These were driven off, only to be succeeded by a further attack by forces from Connaught. These were driven off by a well-timed sortie in the ninth week. Their King, Roderick, was in his bath and fled naked. Finally O'Rourke appeared with a force from Meath, but this was repulsed with heavy losses. The lesson from these enterprises is that if one wishes to be a besieger it is best to be sure one's forces are stronger than those of the besieged; if not he will sally from his castle or town and rout them in open battle.

Henry, who had at first been contemptuous of Strongbow, now began to feel concerned, and decided to go over to Ireland himself. In the meantime Strongbow returned to England, handed over his conquests to Henry, and was relieved to be allowed to be their tenant-in-chief. In this capacity he returned to Ireland with Henry.

But Henry's visit to Ireland was too brief to have any lasting results; he was recalled to deal with a rebellion in his own dominions, and the Irish conquest was not fully consolidated. The Normans had, of course, built castles but many of these were subsequently burnt. The Irish, although incapable of uniting to repulse their conquerors, were at least able to nullify and absorb them.

The event that brought Henry back from Ireland was organized by his treacherous wife and ungrateful sons. The eldest of them Henry, aged 19, was the prime architect; Geoffrey, Richard, and John were to play a fuller part later. The alleged grievance was that each had not been given a suitable piece of territory to rule in his own right, a situation that recalls the attitude of Duke Robert of Normandy to his father. Their claims, although supported by a shadow of French law, were extraordinary and extortionate.

The conspiracy was widespread, and involved baronial risings in England, Brittany, Anjou, and Poitou, an invasion of Northumbria by the King of Scotland, and the conquest of Normandy by young Henry and 20,000 mercenaries. Henry II was more than a match for the situation. While he dealt with the situation in France his Justiciar defeated the southern rebels at Fornham in Suffolk. After that the rebel castle-holders gave little resistance. Perhaps the most surprising event in the whole rebellion was the capture of William the Lion, King of Scotland, when he was jousting in a meadow at Alnwick.

Although the barons' revolt was put down firmly Henry's personal troubles were not over, and never would be. When not jealous of their father, his restless sons were jealous of each other; in 1183 their quarrels involved him in military action once more. Ultimately this filial ingratitude worried him into ill-health and he died at 56, having lost the will to live.

In Henry II the Norman Conquest justified itself. His achievement in establishing and maintaining law and order, his vigour, drive, and wisdom, make him in spite of his faults one of England's outstanding kings. He was brilliant at war but preferred peace, moderate in taking pleasure, but generous in conferring it. Peter of Blois gives a full report of life around Henry. He never sat down except when he was eating or riding. He covered five times as much ground in a day as most people did, and his attendants were constantly on the run. 'No one was more gentle to the distressed, more affable to the poor, more overbearing to the proud. It has always, indeed, been his study, by a certain carriage of himself like a deity, to put down the insolent, to encourage the oppressed, and to repress the swellings of pride by continual and deadly persecution.'

His contribution to castle-building was to use good quality stone, instead of the first material that came to hand, and he introduced buttresses and projecting towers on the curtain walls. He spent £6000 on Dover castle, replacing its wooden walls with stone. He also made some experiments with polygonal design at Orford in Suffolk. The idea of the polygon was that there should be no vulnerable corners, but the early efforts at polygonal castles had too few projecting towers with the result that sections of wall were left unprotected. Under the influence of Eleanor he introduced a few residential improvements into some of his castles, the most noteworthy being at Winchester, Nottingham, and Windsor.

Wallingford appears to have attracted his affection, but Crowmarsh, where he had been humiliated in Stephen's reign, received harsh treatment. When granting a charter to Wallingford in 1155 he prohibited the market at Crowmarsh; Wallingford had of course loyally supported his mother. There had probably been considerable rivalry between the two settlements as the Icknield Way and Grimsdyke (the boundary of Wessex and Mercia) passes through the latter and the Reading to Oxford road through the former.

Sieges were not, of course, always a matter of central politics; very often they were the focal point of local rivalries. In 1158 Ivor Bach of Serghennyd carried out a daring raid on Cardiff Castle. He climbed in by ladder one dark night and carried off Earl William of Gloucester and his wife and son. His motive was ransom. The Welsh had a notable measure of success during Henry's reign. Both Rhuddlan and Prestatyn were destroyed by Welsh insurgents in 1166 after a siege lasting three months.

The same year also saw an unusual siege at Ludlow. It was the most important castle of the Welsh border, being bigger than Chepstow and commanding an entire district. It had been given by William I to Osbern Fitz-Richard, son of Richard Fitz-Scrob of Richard's Castle, which is only four miles away. Osbern passed it to the de Lacys, who strengthened it with a wide and deep ditch. About 1135 the King claimed it because the next heir was not of the direct de Lacy line. In the 1160s the holder was one Josse de Dinant. At this time Warine de Metz, lord of Abberbury, had sent his son Fulk as a page to Josse. The Lacys meanwhile made various attempts to retake the castle or molest its inhabitants. On one occasion a minor battle was fought in front of the walls. During the fighting Josse was attacked and felled by de Lacy and three other knights. Young Fulk, who was too young to take part, was watching from a tower; seeing his master's predicament he seized an axe and rushed to join in. Josse was in dire straits but he had already wounded two of his attackers. Young Fulk whirled his axe to great purpose and, aided perhaps by surprise, killed two and captured two.

Unfortunately for Josse one of the prisoners was Arnold de Lisle. In the castle was Marion de Bruyère who was being brought up by Josse's family. She soon fell in love with de Lisle, who persuaded her to get him a rope of knotted linen. With this he and his fellow prisoner, Walter de Lacy, had no difficulty in making their escape.

Eventually the claimants to the castle were reconciled and Josse went off on an expedition, imagining that it was safe to do so.

Marion then sent de Lisle a message to say she was virtually alone, and he could visit her in safety. She arranged to let down a ladder from the window by which he had previously escaped. De Lisle was fonder of de Lacy and power than he was of Marion so he left the ladder hanging when he went to her room; after a short interval about a hundred men-at-arms climbed the ladder, and spread through the castle. At the given moment the garrison and staff were slaughtered. Roused by screams Marion realized de Lisle's treachery and ran him through with his own sword, which was lying on a table by her bed. Then she leapt from the window and was killed on the rocks below.

Although this episode gave the Lacys possession of the castle it was not long before Josse was back and besieging it. His forces burst their way through the outer ward and were about to secure victory when de Lacy appealed to a Welsh neighbour whom he offered to reward with extensive grants of land. The Welsh attacked Josse in the rear and captured him. Fulk escaped and appealed to King Henry. Henry ordered de Lacy to free Josse, and to evict his Welsh assistants. The latter was easier said than done and took over four years of concerted effort. As Josse died soon afterwards without an heir the Lacys kept Ludlow. Subsequently they probably regretted the efforts they had made to acquire it, for it brought little joy to them and after an interval passed to the Mortimers.

Etienne de Rouen gives an interesting account of Henry's siege of Chaumont-en-Vexin. He sent his Welsh mercenaries swimming into the town down the river. Once inside they fired the buildings. Meanwhile he approached the walls as if to attack. The French moved out to meet him but beat a hasty retreat when the town behind them was in flames. The general confusion enabled Henry to enter and capture the city.

Henry was not a paragon of virtue. He is rightly criticized for raising Becket to such prominence that his power was almost as great as that of the Crown; he did not make much attempt to restrain his quarrelling legitimate sons; of his illegitimate children, reported to number as many as 82, he raised one, Geoffrey, to be Bishop of Lincoln. Geoffrey received the revenues but neglected the duties for fifteen years. His mother was apparently a common prostitute. Henry also apparently produced a child from the girl betrothed to his son Richard (Coeur-de-lion), which traumatic experience may have contributed to the latter's perverted sexual tastes.

However, if Henry II had not ruled efficiently and justly it is unlikely that his son's reign could have existed at all. For Richard spent only eight months of his ten years of kingship in this country. When he returned it was merely to ask the Exchequer to give him more money. Apart from this he sold everything that would sell to anyone who would buy it. William of Scotland was released from his homage to England for 10,000 marks (£6666) and the post of Justiciar was sold for £3000 to a most unsuitable holder. England meant nothing to him except as a means of gratifying his ambitions. He did not know the people and he could not speak the language.

The fact that he was fantastically strong and spent most of his life fighting in a cause that was believed to be romantic has left a false impression of Richard's character. However, in the field of siege warfare he was a figure of very great importance, for he brought back to Europe the experience and secrets of oriental fortification, and, in modifying the accepted designs, incorporated some ideas of his own that made one castle at least far ahead of its time.

When Henry died in 1189 Richard was 32. It took him a year to collect a sum sufficient to launch his Crusade. This done he spent nine months getting as far as Rhodes, by which time he was on moderately bad terms with most of his allies. The fact that the Turks now held Jerusalem and the Holy Sepulchre was in their hands did not have much effect on the rate of progress. But Richard's army, though slow, was methodical and held together. It seems that he was especially severe on food-profiteering, and that he regulated gambling, no knight or squire being allowed to lose more than twenty shillings in twenty-four hours. On his way to the Holy Land he captured Cyprus and married Berengaria of Navarre in the chapel of Limassol. He also fought a sea battle in which he sank a Saracen supply ship.

In Sicily he became so angry with the local inhabitants that his insane rage frightened even his friends according to Richard of Devizes. However, he was not so angry that he forgot how to conduct a siege. When he assaulted their main position he allowed them to fire off all their arrows while holding his own. When the enemy ammunition was exhausted he poured in a devastating stream of arrows and darts, not to mention javelins; the walls were left without guards as no one could look out without getting an arrow in his eye (*quin in ictu oculi sagittam haberet in oculo*). After that the entry was easy, the whole battle occupying only five hours. Nevertheless, he built a massive wooden tower to dominate Messina as a safeguard of good behaviour. Subsequently he took this tower with him to the siege of Acre. This account is taken from the Chronicle of Richard of Devizes, who was a monk at Winchester. Although the Chronicle adds little to our knowledge of sieges, it adds considerably to our information about those who fought in them. In France, a man wishing to seem better than his forbears built a castle in a previously unfortified place. The local people were so incensed they flattened the castle and dismembered the builder. Before Messina, King Richard laid down a law that any foot-soldier who ran away would have a foot chopped off, and any knight who ran away would be reduced to the ranks.

But he is at his best and most topical in describing London: 'Whatever is evil or malicious in the world you will find in that city. Do not mingle with the hordes of pimps; do not mix with the crowds in low eating places; avoid dice, gambling, theatres and taverns. There are more swaggerers there than in all France, and the number of parasites is beyond counting. Actors, jesters, beardless boys, flatterers, men like women, women like men, singing and dancing girls, night wanderers, beggars, mimics, buffoons; this collection fills all the houses.'

He has withering, though different, comments to make about Oxford (an ill-provided place where men and animals eat the same food), Bath (deposited in the depth of valleys in thick and sulphurous vapour, is at the gates of Hell); Worcester, Chester, and Hereford are too close to the deadly Welsh, York is full of dirty and untrustworthy Scots, Ely reeks of the surrounding marshes, Bristol has no one in it who is not, or has not been, a soap-manufacturer. And so on. Outside the city the people are *rudes et rusticos* which sounds slightly better than its meaning of 'coarse and boorish'. Similarly sharp but not bitter comments are applied to people and events. He is cynical but not bitter, satirical but not sour; he can make fun of abuses or pretensions that made other men weep or rage. This was the background to battle, siege, and crusade.

When Richard landed at Acre the city had already been besieged for two years but, in that peculiar way that we have seen before, the besiegers were themselves also partly besieged. Both Richard and Philip of France rapidly succumbed to a fever, doubtless malaria, but Richard was the first to recover. Except when his illness was in an acute stage he was encouraging the Crusaders by deed and word. The main attacks were on a point known as the Accursed Tower. Although the Crusaders poured fire from a large malvoisin tower, the Turks frequently put it out of action by an even taller structure erected on the walls. In the mangonels Richard used large, flint-hard stones specially brought from Messina. It was said that one of these splintered so violently on impact that it killed twelve Turks. Stories like this, of lucky freak shots, were not uncommon. At Tortona in 1155 a similar missile had killed three knights in full armour.

Mines and countermines were all around the walls, but the French eventually reached a position where the wall could be brought down. The defenders of Acre, being short of food, and knowing their danger, appealed to Saladin for relief; his response was to attack the Crusaders in the rear just as they were tackling the walls. Unfortunately for the storming party, the French mine proved insufficient to breach the walls, and the wall-scaling that was meant to accompany it was a miserable failure. But for the garrison enough was enough. The next assault would clearly break through, and the exhausted defenders knew what would happen to them if that occurred. After some negotiation it was decided that the city should be surrendered.

The terms were that Saladin should hand over the Holy Cross and pay a large ransom for the hostages, who numbered two thousand seven hundred. But as time went on it became clear that Saladin had no intention of fulfilling the treaty obligations. He gave neither Cross nor money. Richard could not convey the Turkish prisoners to Jerusalem, and dared not set them free; therefore he marched them out bound on to the plain in front of Saladin's army and massacred them with sword and spear. The Turks could hardly complain for they had forced

Richard's hand; but the cold-blooded massacre was scarcely the sort of action for a devout Christian bent on recapturing the Holy Land from the barbarous heathen. Saladin, on the other hand, had no special reason to refrain from slaughtering the 1500 Christian prisoners he held; and he beheaded them all.

As Philip of France had gone home after Acre, Richard adopted the only possible course, which was to move slowly down the coast towards Jerusalem. His army was reduced to thirty thousand men, and it was soon obvious that European armour left them at a great disadvantage against the lighter, skirmishing Saracens. Another handicap became apparent during the advance. The march was slowed to a snail's pace by the need to transport cumbersome siege materials.

Harassing tactics did not, indeed, have much effect on the Crusaders, and the Turks resolved to bring them to battle at the Wood of Arsouf. To the surprise and disappointment of the attackers, the Crusaders were less weary than had been anticipated. Furthermore the Turks were no match for the fury of the cavalry charges, and 7000 of them were reported left for dead on the field. And so the Crusaders completed their journey to Jaffa.

But Jaffa was not Jerusalem. Trouble was now afoot in Richard's army. As winter came on, cold proved even more devastating than heat had been. Many of his followers had used up all their money, resources, and enthusiasm for the cause.

But the tactical situation was what finally decided Richard. Morale he knew he could restore, but his numbers were too small for an attack on a city as large and well-defended as Jerusalem—or so he thought. Saladin had one army deployed in the hills and Richard knew that if he spread his web too thinly around the walls it would be chopped to pieces by attacks from the rear combined with sallies from the front. Had he been able to bring the hill army to battle his expedition might have ended differently but, not knowing that it was in even greater trouble than his own, he made the wrong decision and postponed his attack. He fell back to Ascalon, in snow, hail, and a mood of growing despair.

From that point the Crusade fell to pieces. Further fighting was seen at Jaffa where the Turks made a spirited but unsuccessful attempt to recover the town. Finally, racked by malaria, Richard was glad to sign a treaty resigning Ascalon in return for the peaceful possession of the other ports.

Richard's return journey, when with alternate caution and reckless folly, he began to thread his way home through hostile territory, is a well-known, over-romanticized story. Suffice to say that he was eventually ransomed for an enormous sum and returned to England in 1194 to find a civil war organized by his brother John; in Normandy his territories were being attacked by Philip of France. The former was soon dealt with, although his possession of Nottingham and Tickhill castles delayed matters for a time. John fled, and Richard only stayed in England long enough to raise the finance for the bitter campaign he wished to launch against his old enemy Philip of France. As the war dragged on for the remaining six years of Richard's life the first payment was not the last, and England was used as a milch-cow for a long futile campaign.

But in the remaining years of his life Richard, whose only interest was in fighting and all that went with it, contributed an outstanding development to siege strategy. This was the famous Château Gaillard, the 'saucy castle' of which we shall have more to say later. For the moment we trace out the rest of his life which ended, as it had always been, in reckless fighting.

In 1199 Richard heard that the Viscount of Limoges had discovered a treasure of Roman gold. The actual finder was a peasant, but doubtless his claims were put on one side in an age when might was right. As the territory lay in Aquitaine Richard claimed it as his feudal right. The Viscount thought otherwise, and decided to fight for its possession.

As a siege it did not amount to much. The garrison of Chaluz Castle numbered only forty. It was not worth a siege engine; all that was required was a breach in the walls and it would be over. While the sappers dug a way to the foundations of the wall Richard amused himself by riding round the castle, offering himself as a target.

In his foolhardiness he was matched by a defender called Bertrand de Gourdon. This man took special delight in exposing himself on the battlements, and varied the occasional crossbow shot with antics on the battlements, dodging arrows aimed at him. But Gourdon was more purposeful than Richard. When the King rode within easy range he took careful aim and drove an arrow straight into Richard's neck.

It was a hazard of war, and Richard accepted it. He had done much the same to others on more than one occasion. It was not particularly serious; he had been wounded before and worse, but he was alive to-day.

But to-day was not Richard's day. The wound festered, and gangrene set in. By the time the castle had fallen he was a dying man.

On his death-bed he had brought to him the man who had aimed the fatal arrow. He had already ordered that the rest of the garrison should be hanged but he wanted to meet his killer. The man was brought to him. The supreme arrogance of the feudal overlord came out in his question: 'Why did you kill me?'—'Because you killed my father and my two brothers' came the reply. Richard was moved 'Set him free,' he said 'and give him a hundred shillings.' Mercadet, the captain of his mercenaries, nodded agreement, and Gourdon was taken away. As soon as Richard died, Gourdon was skinned alive and hanged. (The story comes from William of Newburgh.) Supporting the principle that violence breeds violence, we note that Richard's illegitimate son Philip murdered the Viscount on the grounds that his action in not surrendering the treasure had been the cause of Richard's death.

Richard was a faithless son and a useless king; by some accounts he had unpleasant vices, but he was a warrior, and a man of outstanding courage and skill. As a soldier, his supreme architectural achievement, in which he showed intelligence and artistry, was the Château Gaillard.

This castle, set on a spur of high ground overhanging Andelys, guarded the approach to Rouen. Strategically and tactically it was a masterpiece of siting. It incorporated the lessons he had learnt from century-old, middle-eastern fortifications and he modified them with ideas of his own. In consequence Gaillard had the first stone machicolations to be seen in Western Europe, oblique surfaces to deflect enemy missiles, and deep battered plinths (outward sloping bases) to strengthen the lower walls. It acquired the reputation of being revolutionary in design but this view has been revised to an opinion that it combined the best of the old with some good new ideas. A

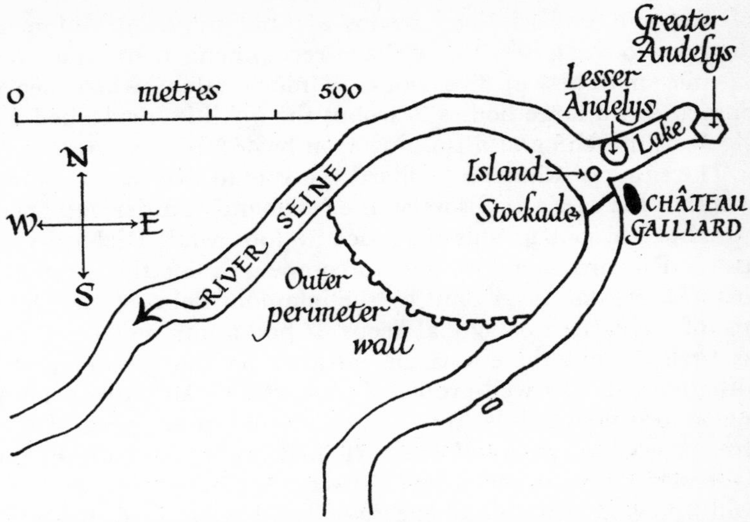

Figure 18.

fourteenth-century writer, John Brampton, is responsible for the story that it took one year only to construct; in fact it took three and even then the keep was not complete. The error appears to have arisen from a remark attributed to Richard: 'Behold what a beautiful daughter of one year'—which he may well have made at the end of a year's progress.

Richard's military bible was the *De Re Militari* of Flavius Vegetius Renatus, usually known as Vegetius. The exact dates of this writer are unknown but he wrote in the declining years of the Roman Empire. The latter part of his book deals with the construction and defence of fortifications, and was highly valued by everyone except the Romans for whom he wrote. Henry II and Richard I are said to have carried it everywhere.

Many of its maxims are famous. 'If you wish for peace prepare to fight for it.' 'Discipline is better than large numbers.' 'Men are seldom born brave but they acquire courage through training and discipline.' 'A handful of men inured to war proceed to certain victory, while on the contrary numerous armies of raw and undisciplined troops are but multitudes of men dragged to slaughter.' Part of the recommended training was 24-mile marches in five hours. Unfortunately when sieges pinned down large bodies of troops the qualities developed by vigorous marching and training soon began to deteriorate.

The siting of Château Gaillard was due to two reasons. One was that the castle at Gisors on the Normandy border had been ceded to Philip Augustus of France by John while Richard was away; Richard therefore had to compensate for this strategic loss. (Gisors had been built by the infamous Belesme, and had one of the earliest octagonal keeps. It had a unique feature for its time in that there was an entrance on the ground floor, although this was well-protected. A curtain wall built in 1123 introduced a principle from Vegetius—the importance of fire from flanking towers. Henry II had built two additional towers. Situated on the border between Normandy and France, controlling the passage of the river Epte, it was a formidable obstacle.) The second point was that the site commanded the approaches to Rouen along the line of the Seine, and therefore blocked the main avenue of invasion from France. The castle stood at the end of a long promontory, was 300 feet above the Seine, and had deep valleys on two sides and a precipitous slope on the third. The site commanded the entire amphitheatre created by the valley formation, and particularly dominated the towns of Great and Little Andelys.

The castle was built in the shape of a ship with the prow facing along the peninsula to a point where the latter narrowed. The outer ward was shaped like an isosceles triangle with three towers guarding the front, one at the point, the other two flanking. At the rear corners were two more towers. Between this and the middle ward was a ditch 30 feet wide and 20 feet deep, crossed by a zigzag causeway. The zigzag in the bridge of the causeway was meant to frustrate the bringing-up of siege engines. Leaving a permanent bridge proved a disaster in the subsequent siege. However, before we leave the outer ward it should be noted that the towers projected unusually far beyond the wall and were set very close together. Objects dropped from the machicolations would ricochet off the buttresses at the base. The forward towers were 35 feet in diameter, and had walls 11 feet thick. The curtain wall was 30 feet high and varied in

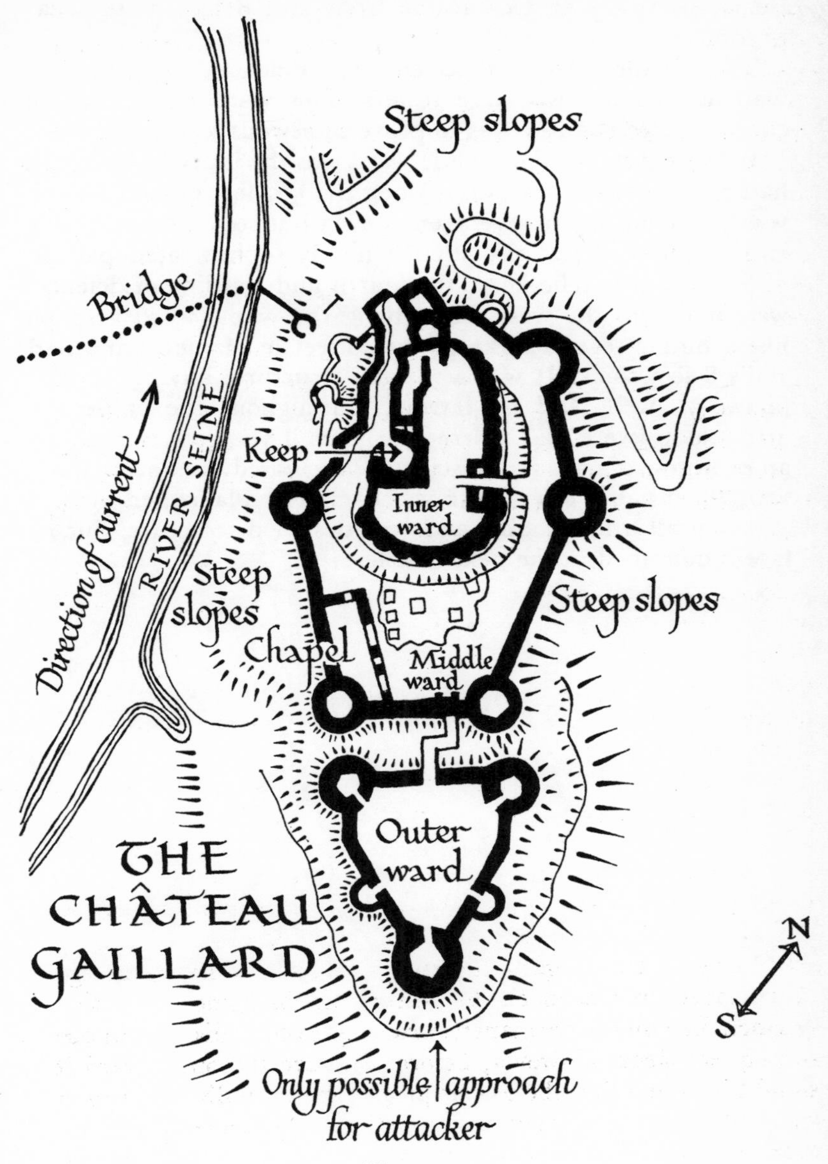
Steep slopes
Bridge
Direction of current
RIVER SEINE
Keep
Inner ward
Steep slopes
Chapel
Middle ward
Steep slopes
Outer ward
THE CHÂTEAU GAILLARD
N
S
Only possible approach for attacker

Figure 19.

thickness, being 12 feet at the front and 8 feet in the rear portion.

The middle ward had its entrance defended by a curtain wall flanked by two large towers. This ward contained the chapel whose window was to prove disastrous to the defence.

In the far end of the middle ward was the inner ward which had a peculiar curtain wall. Along the 500 feet of wall (which was 8 feet thick) are seventeen convex buttresses, covering the entire length. These provided a highly sophisticated pattern of flanking fire. The keep, the fourth and last line of defence, was built into the west wall, giving the whole ward a shape like a human ear. The keep was 48 feet in diameter and had walls 8 feet thick. It was a spartan structure, having no fireplace, no well, and no latrine. Throughout the castle the arrow loops were placed irregularly, as it was recognized that an even row of gaps in the structure increased the hazards to a wall when half a ton of rock was likely to be launched at it.

All in all a formidable and well-designed structure, but its fate is described in the next chapter.

* 7 *

The Small Gains and Large Losses of John

(1199–1216)

AS Richard had no legitimate children the Great Council had the duty of choosing between his elder brother's son, and his younger brother, John. As the former—Arthur—was only 12, John was a unanimous choice in England; the situation was otherwise in Normandy and the other Plantagenet domains, where the thought of an inexperienced boy as king appealed to the barons. When John was imposed on them they rose in spontaneous, though unorganized, civil war. Philip of France extended the war he was already carrying on against Richard's England to include that of his brother; the pretext was supporting Arthur's claims.

John has been severely condemned as a thoroughly bad, mean, incompetent, lecherous king, but this view is challenged by certain modern historians. He had, indeed, numerous faults, but they were shared by other Norman and Angevin kings without attracting much censure from history. As an administrator he was second only to his father; as a military commander he was his equal.

In the early part of his troubles John had the support of his mother, the elderly Eleanor of Aquitaine, and without this he might have lost control of his French territories. It might perhaps have been better for him if this had happened sooner than later. Had he not been in Aquitaine he would not have set eyes on Isabella of Angouleme, who was betrothed to one of his supporters, the Count of La Marche. The latter appealed to Philip of France who had just made peace with John but was only too happy to break it again. Philip adopted Arthur's cause, the barons renewed their disruptive efforts, and it appeared as if chaos would become a permanent condition in John's overseas territories.

A siege that was insignificant militarily, but of far-reaching importance politically, took place at Mirebeau in Poitou, where the old Queen Eleanor was living. Arthur was given the invidious task of besieging and capturing his grandmother, an event designed to give him a little prestige. Eleanor, however, upset the plan by retiring to the keep and refusing to surrender. Even worse, she managed to send a message to John informing him of her plight. John, displaying more military zeal than was his custom in France, dashed off to Mirebeau and after a brisk engagement relieved the castle. In the course of this he captured Arthur.

It is unlikely that John had any strong feelings about his nephew but there was no doubt that he had to regard him as a danger. It is said that he tried to persuade Arthur to break his friendship with Philip of France, but was laughed at. Arthur was put in a dungeon of Rouen castle and was never seen again. Speculation about the manner of his death has been continuous, but the most likely account describes how he was stabbed and thrown into the Seine. The year was 1203 and Arthur was sixteen.

The murder was not discovered for several months but when the news leaked out there was uproar. Arthur's former supporters transferred their allegiance to Philip of France who took the melodramatic, and futile, step of summoning John to Paris to stand trial for murder. When this summons was ignored John was tried in his absence, found guilty, and declared to have forfeited all his possessions in France.

The war which was the natural consequence of these events was half-hearted and inept. Philip invaded Normandy and began some lengthy sieges; John crossed from England but made no attempt to engage his enemy, or pursue any sensible military policy. After a series of encouraging successes against minor fortresses the French were emboldened to take on the great prize that was the key to Normandy—Richard's Château Gaillard.

The layout and construction of the castle itself has already been described in the previous chapter. Details of the siege come from Guillaume le Breton, the French King's chaplain, who was an eye-witness.* Guillaume le Breton firmly believed that

* His account contains one or two errors that are apparent in an examination of the site.

John had murdered Arthur, but it is doubtful whether this conviction was widely shared or had much influence on the fighting.

Philip Augustus had a formidable task. Like Richard I he was a keen student of Vegetius and was therefore well aware that one of the secrets of success is careful preparation. He arrived with his army in August 1203 and spent the first month weighing up the situation. The weakness of castles, as we have already noted, is that they can be starved out. In order to prevent this happening to Château Gaillard Richard had fortified a small island in the middle of the river with an octagonal tower, palisades and ditches, and blocked the passage of the river by a triple row of piles. The island was linked to the mainland with a wooden bridge, which was in turn protected by towers on the castle slopes. In theory this should have denied the passage of the river to all but friendly shipping. In practice it did not, for the French army promptly destroyed the bridge and sent a posse of strong swimmers to cut the palisade. These pioneer frogmen were assisted by a diversionary attack staged on the island fortress.

Fortunately for the French, John was going through one of his periodic attacks of lassitude. Had he displayed the military vigour which he undoubtedly possessed it is most unlikely that the French army would have crossed the river without sustaining heavy casualties. By the time John reacted to the situation the French were on the right bank and encamped under the walls of Little Andelys. The Earl of Pembroke was then given the doubtful pleasure of launching a relief through a French army which was on both banks of the river and was linked by a pontoon.

The English plan was a complicated night operation and appears to have miscalculated such vital factors as the speed of the river current and the hazard of mounting combined operations in the dark. Seventy boatloads of food and weapons were to go up river accompanied by a force capable of assaulting and breaking the defences of the French blockade. While this was going on the French army would be attacked in the rear at the neck of the castle peninsula, and therefore be too occupied to assist the defenders of the boat blockade.

Unfortunately the heavy current slowed down the water-party so much that the land party attacked too soon. The

French were able to concentrate first on one group and then on the other, routing both, and that was the end of the relief. John shrugged his shoulders and departed for England.

But the castellan, Roger de Lacy, was a man of courage and experience and there was a reasonable chance that the besieging army might get more than it bargained for in a full-scale attack. So he reduced the ration scale and waited.

But if de Lacy was fully resolved to hold Gaillard, Philip was equally determined to capture it. He did not allow his army to be whittled away with diversions, as so often happened in sieges; instead he employed it in digging trenches and building towers so that the castle was completely cut off from outside assistance; in February when he judged the moment to be right, he launched a tremendous assault on the outer tower and adjoining curtain. The tower stood the strain but the curtain did not; before long the French were over the ditch and through a breach in the wall. The key to their success was not the massive trebuchet they had built but the insidious work of miners who, working under mantlets, picked out the foundations of the structure, filled the gaps with wood, and then burnt the props. Subsequently John used the same tactics at Rochester.

But the outer ward was by no means the whole castle. The next stage involved crossing thirty feet of ditch and assaulting towers that were built flush with the cliff face. Philip's siege engines began battering away but were obviously not going to effect a breach without assistance. The miners found themselves thwarted by the fact that the lowest foundations of the walls were out of their reach. They tried climbing the sheer side of the ditch by sticking daggers into the chalk and using them as makeshift ladders, but those who reached the walls could not obtain sufficient leverage to break into the stone. Meanwhile Philip was making belfry towers out of unhewn tree trunks.

At this point one of the French soldiers, who probably knew the castle well, observed that a garderobe (latrine) emptied on the west side. Just above this was a chapel window that was not barred as might have been expected. He crawled up this unattractive path, entered the chapel, and pulled in a few companions through the window.

They were now in the crypt, and were unobserved, for no attack had been expected on this side. By a prearranged plan they suddenly raised a tremendous clamour giving the impression of large numbers. The garrison tried to smoke them out by lighting a fire at the entrance, but the smoke blew back in their own faces. In front of the main gate a further tremendous clamour was raised. The garrison already weak through hunger and losses, panicked and retired to the inner ward. The jubilant chapel-party thereupon rushed out and opened the main gate for their friends. But even then the siege was not over. Miners had taken the outer ward, and trickery the middle ward, but the convex construction of the inner ward now made it extremely difficult for miners to get their picks in, and the situation too concentrated for subterfuge to be effective. At last the catapult came into its own. Philip had an enormous machine which he named 'Cabulus', and this began hammering the stonework of the inner keep.

But if the siege of Château Gaillard proves anything it is that no castle is impregnable when the miners can get their picks in. At Gaillard by reason of the convex patterning mentioned earlier their task was exceptionally difficult, but it was aided considerably by the presence of a rock bridge over the ditch. Under the cover that this afforded they were able to chip into the main structure sufficiently far to assist the crushing blows from Cabulus. Rather foolishly the defenders burrowed back from the inside to scare off the French miners, but these tactics failed in their object and further weakened the structure. Finally, a large quantity of masonry toppled over leaving a breach that the defenders were too enfeebled by starvation to hold. Abandoning hope of defending the keep they tried to escape by a postern, only to run into their captors. On March 6th, 1204, the last twenty knights and one hundred and twenty men-at-arms laid down their weapons. The legend of the impregnable château had lasted a mere seven years. Soon after, Falaise, equally strong, fell through treachery.

The dark side of the siege of Château Gaillard resembled the siege of Calais, described in Chapter 10. When Philip surrounded Gaillard the population of Little Andelys retreated into the castle, but when supplies ran short a thousand 'useless mouths' were sent out. These were allowed to go to safety through the besieging forces. A little later a further four hundred were sent out, and included the sick and women and children.

But these were not allowed transit. Instead they were kept in no-man's-land where they lay through the winter eating grass and even practising cannibalism. Finally the French relented and allowed the survivors through, but most were too far gone to be able to recover.

With Gaillard and Falaise gone, John's overseas empire was doomed. Rouen and the remainder of the Norman cities surrendered without serious resistance. Anjou and Touraine soon followed. By 1206 only Bordeaux, La Rochelle, and southern Guienne remained of his great continental dominions.

This tide of misfortune at last stirred John into half-hearted action. He landed in France once more, besieged and took Montauban, and burned Angers. But the effort lost impetus, and before long he returned to England; from henceforth English kings and barons would owe allegiance to one side of the Channel only. John, without knowing it, had conferred an enormous and lasting benefit on this country.

Next John stumbled into a quarrel with the Pope. When the Archbishop of Canterbury died the junior monks elected a successor without reference to the king or their seniors; in this they were encouraged by the arrogant Pope, Innocent III. John refused to acknowledge the monks' choice and substituted one of his own. The Pope ratified neither, but appointed yet another, one Stephen Langton. John reacted by refusing to accept Langton, and would not even allow him to set foot in England. The Pope then laid an Interdict on the country.

An Interdict was an event of enormous and shattering significance—only priests were allowed mass, the dead were buried in unconsecrated ground, marriages took place in churchyards; inside the churches all crosses, relics, statury, and images were placed on the ground. The effect on the people was to make them feel they had been disowned by God. John, on the contrary, was merely scornful of the Interdict, and found it offered a useful opportunity of pocketing money that would otherwise have gone to the Church. In 1213 he was compelled to take the Pope more seriously—not on religious grounds, for it is reported that John had in a temporary whim become a Mohammedan—but because Innocent had empowered the French King to drive him from his throne, blessing the enterprise as an official crusade. Philip of France assembled

1700 ships to transport his army; John assembled 60,000 men at Dover to oppose it. But the morale of the defending force was virtually non-existent, and when John was approached by the Papal legate, Pandulf, the latter was able to portray such a glowing account of the invasion force that John's own position appeared hopeless; in all probability it was. Without consulting his council, who could hardly complain as they were not supporting him, he conceded that England should become a Papal fief and pay 1000 marks a year to the Pope. The effect of this submission was to recognize the Pope as the secular as well as the spiritual overlord of the kingdom. It was an astute move in one sense, for the Pope must now pardon and support John. The French, with a valuable prize in their grasp, saw it wrenched out of their hands. Before long Philip of France had quarrelled with one of his principal supporters, the Count of Flanders, who was already in league with the English King. This division of his enemies enabled John to launch an attack on the French invasion fleet as it lay in the Port of Damme, and to win the first recorded English naval victory.

Encouraged by this success but insufficiently supported by the barons he determined to recapture the territory lost to Philip of France. With a large army of mercenaries, and some dubious allies, he set off on the path of reconquest. But John's house of cards fell to the ground after the defeat of Bouvines in July 1214, and he had to beat a hasty retreat to England.

On his return he decided to chastize the barons for their failure to support him; but this ambition had effects that went far beyond expectations. Prior to Henry II the monarchs had been little more than the first among equals, but through such measures as the Assize of Clarendon Henry had tilted the balance of power sharply in favour of the Crown. By John's time the Crown was strong enough to enforce its will. This fact did not pass unnoticed by the barons, but their reaction to it was vastly different from that taken on previous occasions when they had wished to curb royal power. In 1214 they held a meeting and drew up a Charter of Liberties which they expected the King to respect. And in 1215 John was forced to sign it on the well-known occasion at Runnymede. Theoretically, the Magna Carta might have been the happy culmination of a troubled reign but even the wildest optimist could hardly have expected that. As soon as the barons had returned to their own lands John assembled a mercenary army and set out to discipline them one by one.

This last phase of his reign produced two remarkably interesting sieges. The first was at Rochester, the second at Odiham.

The siege of Rochester opened the campaign.

Well aware of the fact that John was steadily building up an army by bringing in groups of foreign mercenaries, the barons decided they must take the initiative. Although John had won the approval of the Pope by taking the vows of a Crusader, he was still opposed by Stephen Langton, the Archbishop of Canterbury, in whose control lay Rochester Castle. The barons, therefore, decided on a swift occupation of this strategic point and sent William d'Albini to execute the plan. As the castellan opened the gates and welcomed him it seemed a successful move.

Unfortunately for the Rochester garrison the pace of events did not leave time for adequate victualling. John, who had a highly capable force of foreign troops, moved quickly to the siege, which he conducted with ferocious energy. Even the loss by shipwreck of a large reinforcement from overseas did not affect his morale, although it certainly increased his ill-temper. The barons sent a force from London to relieve the garrison but had to beat a hasty and ignominious retreat. The fight continued over seven weeks (Roger of Wendover gives it three months) and was one of the bitterest conflicts of the era. It is clear from the chronicles that this was no ordinary siege but one of those battles which occur from time to time when men fight on until they are all killed or wounded. In the boxing ring this sort of encounter is called 'slogging it out toe to toe', and occurs when men abandon all their skill and science, and simply fight till one drops. Wendover compares the Rochester fighters to wounded lions. For every stone that was launched into the castle one went back, for every arrow or dart one would be returned. John nearly suffered the fate of his brother Richard for he once came within easy crossbow range of the castle; the best archer thereupon asked permission to put an arrow through him. D'Albini refused, saying it was God's task to settle with John, not man's. This clemency was compared with

David sparing Saul's life. The incident illustrates the peculiar protocol which pervaded mediaeval warfare, and explains the surprise Richard I felt when he was wounded by a common archer.

But Rochester, like Gaillard, fell to the sapper not the gunner, although, in modern parlance, it was certainly softened up by the latter. First the sappers broke through the curtain wall, and a desperate fight took place in the bailey; then the garrison retired to the keep. To all intents and purposes the battle was won and starvation would have settled the final issue but John would have none of that. Instead he set the miners to work on the south-east corner of the keep. It is possible that this mine tunnel had already begun outside the curtain and was well on the way to the keep when the curtain was breached. Be that as it may, a substantial chamber was soon in being underneath the corner of the keep. According to the usual custom it was propped with beams and filled with brushwood, but an additional refinement was to pack it with the carcases of forty fat pigs. When these were set alight the heat was so intense that the foundation cracked and the tower and a portion of the wall fell outwards (Plate 7). Even then the garrison fought on, for the architects had ingeniously divided the keep into two halves with a cross-wall. But eventually the effects of starvation and exhaustion could not be set on one side, and Rochester surrendered. John ordered William d'Albini to be hanged but was dissuaded from this course by one of his mercenary captains, Savaric de Mauleon, on the grounds that the ensuing retaliation elsewhere might be a problem. The outcome was that the knights were imprisoned in Nottingham and Corfe castles for a time and the remainder of the garrison was summarily executed, including the archer who had been persuaded to spare John's life.

Rochester was a disastrous start to the baronial campaign. Soon after this humiliating blow they received another. John persuaded the Pope to excommunicate them for rebelling against one who had taken the Cross—a penalty, in that superstitious age, which by most was more feared than physical death.

The King followed his capture of Rochester by a devastating sweep as far north as Berwick. This harrying expedition was intended partly to impress the rebel barons and partly to provide an occupation for John's mercenaries, who, if not given a task, might choose a less suitable one and execute it thoroughly. But the effect on the barons was less to intimidate than to make them acutely aware of their extreme peril. Their reaction, like that of most frightened people, was not a wise one. They declared John deposed and elected in his place Prince Louis, son of King Philip of France. It was a desperate and stupid move. France was already identified as the national enemy and it would have been difficult to find a less popular choice than Louis. This fact was soon brought home to him.

He failed to capture Dover which his father had warned him was vital to the success of his whole campaign. Instead the garrison beat back his assaults with such vigour that he was glad to withdraw his forces out of range. Having failed at Dover, he had more success as he moved inland and was able to take Reigate, Guildford, Farnham, Winchester, Odiham, Marlborough, and Worcester. Of these Odiham was the most notable. Only the ruins of a single tower now remain of this castle, which is on the edge of the Whitewater river, a few miles from Basingstoke. Even to-day the marshiness of the ground makes it difficult to approach, and in its heyday it was doubtless as formidable as any other fortification in a watery setting. It was, indeed, a popular royal residence on account of the surrounding hunting; the adjoining forest was comparable with Windsor. In the Pipe Rolls for 1207 John paid 5s. for a wolf caught nearby and 6s. for the heads of six Welshmen. Who the Welshmen were and why their heads should have been sent to Odiham is not known. A rather more congenial arrival was twenty tuns (5040 gallons) of wine which John had ordered to be sent to Odiham on April 15th, 1216, although he did not stay to drink it. It may well have helped to give heart to the garrison for when the French army arrived the castle held out for fifteen days although its fighting strength numbered only thirteen—three knights and ten soldiers.

Louis was undoubtedly short of siege artillery but he already had a number of conquests to his credit, including the powerful castle at Marlborough, so that Odiham must rank as one of the most skilful and courageous defences in history. The slender garrison did not merely hide behind the defences but made frequent sallies. On one of these occasions it took back thirteen prisoners. Undoubtedly the main strength of Odiham lay in its site, and the fact that the garrison knew the pathways through the morass. However, some writers hold the view that Odiham demonstrates the military excellence of the juliet, which is the name given to circular keeps, such as Conisborough (Yorks),

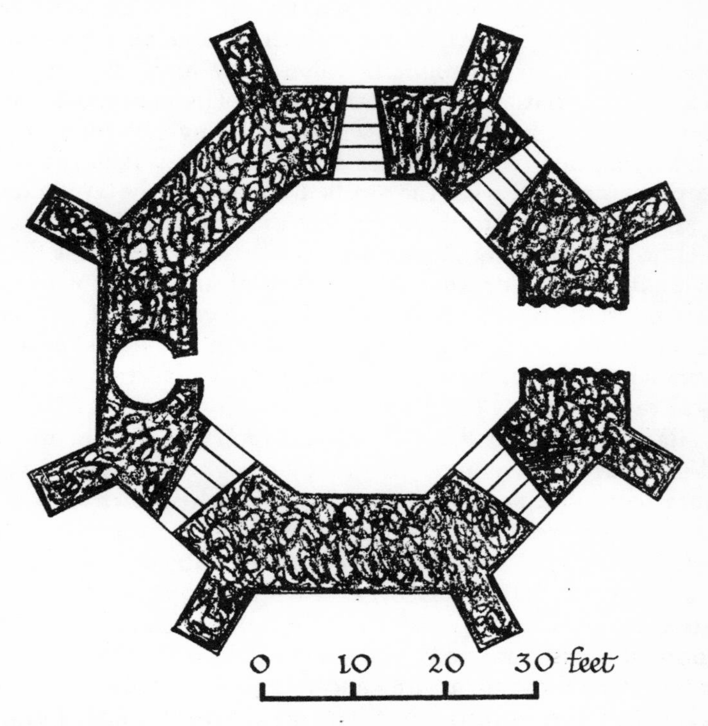

Figure 20. Note the thickness of the walls. The height of the tower was about 65 feet.

Orford (Suffolk), Skenfrith (Monmouthshire), and Launceston (Cornwall). For certain purposes, juliets were excellent fortifications. They could be manned by small garrisons, offered a deflecting target to siege artillery, and were not encumbered by outworks within the field of fire. They were ideal for frontier posts where they could delay armies, and they appear later on the Scottish border as 'pele' or 'peel' towers, often enclosed in a single wall known as a 'barmkin'. But there was a vast difference between the military potential of juliets or peels and the power of a large castle such as Lewes or Rochester.

Hearing of Louis' setbacks, John decided the time might well be right for a decisive battle. He took Lincoln, but while marching south was caught by an unduly high tide at Fossdike, on the southern side of the Wash. His march had been a gambler's throw that went wrong. Although he himself was already in safety when disaster occurred, the sudden return of the tide combined with the swollen current of the Welland to create a whirlpool; in this John lost the sinews of his army.

Although he may have been poisoned, the most likely explanation of his death is that he died of some fever which, in his exhausted state, he was not able to overcome. It has been suggested that he died of dysentery brought on by exhaustion. Dysentery is not, of course, brought on by exhaustion, nor does it kill rapidly.

Thanks to Magna Carta an impression has grown up that John's barons were men of great foresight and sagacity. A somewhat more accurate view emerges when their activities and amusements are examined. Watson in *Memoirs of the Ancient Earls of Warren and Surrey* (1782) a book in which he tries to prove that Sir George Warren was true heir to the Earldom of Warenne, gives a story of the sixth Earl, who 'standing upon his castle walls in Stanford, viewing the far prospect of the river, and meadows under the same, saw two bulls fighting for a cow. A butcher of the town, owner of one of the bulls, coming accidentally by with a mastiff dog forced his own bull into the town by means of the same, who was no sooner entered but all the dogs of the town joined in the pursuit; and the bull being thus mad with the noise of the people, and the fierceness of the dogs, overturned everything in his way, which, causing the inhabitants to rise in a tumultuous manner the earl on hearing the noise mounted his horse and riding into the town was so pleased with what he saw that he gave all those meadows, in which the two bulls were first found fighting (since called the castle meadows) perpetually as a common to the butchers of the town (after the first grass is eaten) to keep their cattle in till the time of slaughter, on condition that they found yearly for ever, the day six weeks before Christmas (being the day on which the sport first began) a mad bull for the continuance of that diversion. A custom which I am informed is kept up to the present time.'

This earl had a natural son named Griffin. Although illegitimacy in mediaeval times was not a social handicap it carried certain material disadvantages. Children had to vary the colours on the family coat-of-arms (in Griffin's coat the gold and azure became silver and black) and they tended to receive the less salubrious properties as their share of the estate. Of Griffin's castle Watson says, 'In one of the Harleian MSS 2131 it is said that in the county of Salop, two miles from Ichtefeld was an ancient castle, situated on a terrible morass, by a river side, which in times past was inhabited by the Earls of Warren and Surrey, and was called Earl Warren's castle.' The 'terrible morass' must have made it a magnificent defensive position but it cannot have been a comfortable or cheering residence. Between boredom and the discomfort of living in damp and bleak fortresses it is not perhaps surprising that the barons regarded any unusual or brutal incident as immensely diverting, nor that, when there was no other relief, they would quarrel or rebel without any apparent reason.

❋ 8 ❋

The Long Reign of Henry III

(1216–1272)

THE boy of nine who came to the throne in 1216 was John's eldest son. Unlikely though it seemed at his accession, he was destined to have one of the longest reigns in English history—56 years. In 1216 the political situation was unstable and threatening. Louis' supporters had gone so far in rebellion that they could not easily draw back, and it was doubtful how far the barons would be prepared to support the helpless son of a hated father. But in the event the rebels lost heart, and the majority of the barons remained loyal. Louis' supporters became restive under his arrogance and there was wide-spread resentment at his presumption in granting estates in England to some of his French followers. Matters were finally clinched by two battles. The first was at Lincoln, the second a naval encounter in the Channel.

Lincoln was a powerful royal stronghold, and as such a natural objective for Louis. He collected an army of some six hundred knights and several thousand men, marched briskly north, and allowed his army to commit all the excesses that are particularly hateful when performed by foreign troops. The town offered no opposition to Louis' entry but the castle was bravely and skilfully defended by a heroine of the time, one Nichola de Camville.

The royal forces, under William the Marshall, could not match Louis in numbers, but made up for that in skill. The army was deployed in seven divisions, the bowmen a mile ahead, the baggage a mile behind—an arrangement which deceived the French leaders into thinking that two armies were arriving. Drooping morale was further damaged by a brisk sortie from the castle. In the confusion Louis' army was so trapped in the crowded streets that it could not deploy and could only put up insignificant resistance. Although the usual slaughter of lower ranks occurred only one noble was killed, the Count de La Perche, commander of Louis' forces. Unhorsed in a churchyard he was called upon to surrender. He replied he would never surrender to an English traitor. This remark so incensed his challenger that he speared the Count through the vizor.

On account of its grotesque nature this battle was subsequently nicknamed 'The Fair of Lincoln'. The initial dispositions were extraordinary. De La Perche posted squadrons of defenders in front of each of the four main gates of the town but omitted to do so in front of the castle on the western side. It was, therefore, a simple matter for the royalists to enter in small numbers through the castle postern and emerge into the city. William the Marshall approached this apparently easy ingress with some caution, fearing a trap. He sent Bishop Peter, who had the military skill which has been noted in other bishops of the period. The Bishop decided that the best plan would be to unblock a disused gate as this would enable the relieving troops to enter the town in greater force. As unblocking the gate would draw the defence to the spot, which was lightly defended, William sent Fawkes de Bréauté into the castle with a contingent of crossbowmen where he arranged a sally which was very effective. William's troops followed and quickly passed into the city where the streets became so crowded that at times fighting was scarcely possible. This factor of overcrowded streets earned the battle its nickname. The final incongruity was when the rebels tried to flee through a narrow gate which had a swing door which closed automatically. In the general confusion an alarmed cow was swept along and blocked the gateway, jamming the door as well. In consequence one half of the rebel force was made prisoner, and casualties were much lighter than might have been expected. The unlucky person was the Count de La Perche.

Equally unfortunate was the commander of the French fleet, a celebrated pirate known as Eustace the Monk. His fleet was taken by surprise, and then completely outmanoeuvred. The English first attacked with a flight of arrows, then grappled their own boats to the French with hooks and chains. Quicklime was scattered down wind so that it blinded the French crews; those who avoided this found the English on board hacking down their rigging with axes. Eustace tried to surrender and offered a large ransom but his proposal was scorned and his head perfunctorily chopped off.

With the war lost Louis was content to return to France, the more so as he was given a substantial sum of money to hasten his departure. England was free of the invader but peace was still far from being assured. During Henry's minority the government of the country rested in the capable hands of the Justiciar, Hubert de Burgh. By 1220 Hubert decided that the time had come to move against the disaffected barons remaining from the previous reign. In order to retain their support John had rashly given away crown lands and royal castles; and on his death the holders, some of them foreigners, had refused to hand them back saying they were holding them in trust during the young King's minority. In order to dispose of this dubious claim, Hubert declared Henry of age and demanded that the castles should now be returned.

The request was not popular. The Earl of Aumale defied the royal forces to take Rockingham, and when they did, seized two other castles. A fierce struggle which combined excommunication with brisk military action ended in his banishment.

Bedford was a tougher nut to crack. One of the most detested of John's foreign favourites was Fawkes de Bréauté, the same soldier who had done valiant work at Lincoln. His obduracy brought about the siege of Bedford in the second half of 1224, although in the event he came no nearer to the castle than Northampton. In 1215 the lord of Bedford had been William de Beauchamp who sided with the rebel barons, and admitted them to the castle. In consequence he was besieged by Fawkes de Bréauté and compelled to surrender after seven days, for which service Fawkes was awarded the castle and had the grant confirmed by Henry III. Within a few years of the new reign it became obvious that Fawkes had cast himself for the role of local tyrant. Eventually neither his neighbours nor the King were prepared to stand any more, and one Henry de Braibroc, a judge, was sent to Dunstable to hear complaints. Braibroc gave thirty judgements against Fawkes, each verdict carrying a fine of £100. This may be compared to a contemporary fine of some £200,000. Fawkes laughed scornfully and kidnapped the learned judge, incarcerating him in Bedford Castle in humble circumstances. Henry ordered that Braibroc should be released but Fawkes rejected the demand. From now on Henry prepared carefully to suppress his vassal.

The scope and thoroughness of the siege of Bedford Castle make it one of the most interesting of mediaeval battles. Fortunately, detailed accounts from contemporary sources give a clear picture of events; less fortunately, the thoroughness with which the castle was subsequently razed has obliterated all but a few minor points. However it is possible to make a few deductions from the nature of the site and the course of the siege.

Situated as it was on the left bank of the Ouse on a gravel plain it seems likely that considerable use was made of water defences. However, it is clear that only a part of the structure was protected by water, for miners are mentioned at every stage in the assault.

The opening move was the excommunication of the castle garrison, a service which was performed by the Archbishop of Canterbury in person. Excommunication tended to lower morale as men were less likely to risk their lives if the aftermath was eternal damnation. Nevertheless, there was no weakening of morale at Bedford.

Another preliminary was the assembly of vast quantities of materials from all over England. Records of this are seen in the orders to local sheriffs. They called for men, money (such as unpaid scutage), iron, steel, hides (for protecting engines), leather, quarrells, stones for missiles, masons, miners, carpenters, food and wine (in huge quantities), and even greyhounds to provide sporting diversion. Fifteen thousand crossbow quarrells were sent from Corfe Castle and a further four thousand were ordered to be manufactured at top speed in Northampton. Miners came from as far afield as the Forest of Dean, charcoal came from Gloucester, and knights from distant Lancaster. The local towns were combed for both men and materials; even the bishops had to supply one man for every sixty acres of church land to assist the *gynours* on the siege engines.

The castle was commanded by William de Bréauté. On realizing that the King meant business, Fawkes had departed on a swift tour to rally support, mainly from the Earl of Chester. Although Chester would probably have been glad to assist in a little rebellious disaffection he was unable to do so for he had already been included in the royal force, where he was under close observation.

The siege began on June 22nd and pressure was not relaxed until the garrison surrendered on August 15th. A petraria and two mangonels concentrated continuous fire on a tower on the east wall, two more mangonels battered another tower on the west wall, while mangonels were also ranged on the north and south walls respectively. Two large belfrys overlooked the castle and raked it with a continuous shower of arrows. The cat began to bite into the walls while the slingers made it suicidal for anyone to show his face over the battlements.

The barbican fell with the loss of less than half a dozen men but the outer ward was briskly defended and casualties mounted sharply. The outer ward was, as usual, a general storehouse enclosing not only arms, ammunition, and forage but also horses, sheep, cattle, and pigs. Once this storehouse was lost the garrison knew that their only chance was to fight back with such vigour that the attack would not be able to stand the casualty rate. The theory of a fortified position is that small forces are able to inflict injury much beyond their number. Therefore, once they are jeopardised by failing food or ammunition supplies, the hotter the encounter the better are their chances; for the besieger may unwisely abandon his assets, such as the use of starvation, and fling himself into a close encounter in which the advantage lies with the defence.

The first phase of these tactics worked well for the defence: and the attackers, who were drawn mainly from Dunstable, suffered severe losses. But Bedford Castle was besieged by more than a single barony; it had against it the entire resources of the kingdom. Further casualties occurred among the miners assaulting the wall of the inner bailey but there were plenty of miners. The breach in this wall was also vigorously defended and many lives were lost on both sides. Finally the garrison retired to the old tower on which the concentrated resources of the entire siege were now centred. Once more it was the miner and not the mangonel which did the damage. So thorough was the work that when the mine chamber was fired the walls split and the whole building tottered. Resistance was at an end.

William de Bréauté was hanged, and eighty of the garrison, mainly knights, shared his fate. His wife was freed from blame and allowed to retain a substantial portion of land. Gilbert de Bréauté was allowed to retain one manor and Fawkes' wife was allowed two—subsequently she divorced him. Strangest of all was the fact that Fawkes, the mainspring of the trouble, who had fled to Wales on hearing of the castle's fate, later surrendered at Coventry and was allowed twenty marks as a travelling allowance to exile in Rome. But he complained to Earl Warenne, who escorted him to the ship, that he could not understand why he was treated so harshly.

There are interesting records of the disposal of the siege materials. Nine hundred crossbow quarrells returned to Northampton, as did the siege engines, for storage in the castle. Other materials were shared out as payment, or awards for good service.

And that was the end of Bedford Castle in every way. Five days after the surrender the demolition of the castle began. The banks were used to fill the ditches. Stones from the buildings were to be sent to churches and priories; Braibroc was charged with the task of supervising this. The mound was reduced and the bailey walls, where they still stood, were lowered to half their former height. Subsequently a manor house was built on the fifteen feet high stump of the mound but later this also disappeared and modern times saw this level surface converted into a bowling green. To-day it is impossible to trace the exact shape of that most formidable castle which was originally built by Payn de Beauchamp in the eleventh century, stood two famous sieges, but was too great a threat to be allowed to survive in the unsettled condition of the thirteenth century.

Although Henry officially came of age in 1227 he did not then, or later, show much sign of wisdom or maturity. In 1232 he held Hubert de Burgh responsible for the failure of an expedition to France. The King's retainers removed Hubert from a church in which he had taken refuge, and carried him naked, shackled to a horse, to the Tower of London. After a farcical trial he was committed to Devizes from which he made a memorable escape; on a dark night he climbed over the battlements and dropped into the moat, Eventually, he was reconciled to the King and restored to his estates and honours.

The next twenty-four years of the reign were marked by extravagance and misgovernment in which Henry relied heavily on foreign favourites. By 1258 the mounting discontent of the barons had found a spokesman and leader in Simon de Montfort, Earl of Leicester and brother-in-law to the King. Ironically, the very man who led the oppositon to Henry's foreign advisers was himself an alien and had been in England for only twenty years. He had considerable military experience and was well aware of the hazards of war; his father had been killed by a stone from a mangonel at the siege of Toulouse in 1218.

In 1261 Henry attempted to break the shackles that Parliament was, very rightly, trying to impose on him. Sporadic fighting developed which included an excellent example of a desperate siege in 1263. In most recorded sieges retreat to the last bastion of the keep was the preliminary to defeat, but at Rochester in that year a heroic defence was rewarded by timely relief. The details come from John Watson's *Memoirs of the Ancient Earls of Warren and Surrey.* Earl Warenne was defending Rochester against a force supporting Simon de Montfort. 'The King who was then keeping his Easter at Nottingham, having intelligence of this, marched with great expedition to relieve him; and though the assailants had won the bridge, and the first gate of the castle by assault, and nothing remained to the garrison but one tower, yet the place was so manfully defended that the King came up in time; for the besiegers having notice of his approach, and not daring to give him battle, retired to London, leaving only a few soldiers behind, which were soon after discomfited.'

In 1264 the two sides clashed at the Battle of Lewes, which, although fought near the castle, was not a siege. The royalists were defeated. Both the King and Prince Edward were captured: and Earl Simon, who wisely kept Henry, a reluctant prisoner, under his own immediate control, became for a time the *de facto* ruler of the country.

At one stage young Prince Edward was confined to Wallingford Castle, from which a group of royalist supporters tried to rescue him. They were led by Warine of Bassingbourn and though their numbers were small they managed to break into the outer ward. At that point they were checked. Wallingford was enormous and entrance to the outer ward gave them little more than a nuisance value. It is said that the garrison threatened to deliver Edward to them by mangonel if they did not depart promptly. Even if the rescuers did not believe this threat, Edward appears to have done so, as he sent a message requesting that they should abandon this wild enterprise. However, Warine of Bassingbourn seems to have been a gainer in the long run for he was soon after granted the manor and castle of Astley. (It was rented from the Earl of Warwick for £5. 10. 6 per annum; formerly it had been held by the service of holding the Earl's stirrup whenever he wished to mount a horse.)

Unfortunately for his cause De Montfort did not keep a sufficiently close guard on young Prince Edward, who escaped in 1265 and rallied a force in the west. The future Edward I was not long in showing the tactical brilliance that was later to make him one of England's most successful warrior kings. He surprised Simon the Younger outside Kenilworth and chased him naked into the castle. Then he swung back to intercept the older Simon at Evesham—one of the bloodiest battles of the century. So fierce was the fighting that even the helpless prisoner, King Henry III, was nearly killed by mistake. Earl Simon, who had governed England for fourteen months, was killed, his body mutilated, and his head stuck on a pole.

Although the Battle of Evesham settled the political issues of the war it did not end hostilities. Rebel resistance continued from three main centres: Ely, Axholme, and Kenilworth. All were protected by water or marsh. Ely's qualities have already been noted, but Kenilworth was a newcomer to the scene of desperate and dogged resistance.

The site of Kenilworth, which has been drained since the Civil War, does not look very formidable to-day. The castle buildings stand on the end of a low spur which juts out into meadows. Between the castle grounds and the remainder of the spur is a dry ditch. The keep has an air of massive solidity, having walls twenty feet thick and a solid first storey, but it looks no more impregnable than any other structure of its type and time.

However, in 1265 Kenilworth was surrounded by a lake covering 111 acres. It is obvious that the original holder, Geoffrey de Clinton, had selected the site because it was a hillock in marshy ground; subsequently it became a royal stronghold. By Henry III's reign two streams had been dammed to form the lake, and the castle itself was fully equipped and organized as a central strongpoint and depot. In 1254 Henry granted it to Simon de Montfort, Earl of Leicester. This grant was an act of outstanding folly, for Montfort was already accepted as the leader of the rebel barons, and the fact that he was married to the King's sister was not going to make him any less of a threat to the powers of the throne. Some time before Evesham, Montfort had prepared Kenilworth as a base for dominating the Midlands, and after his death his son was able to hold off a siege as well organized as that of Bedford some thirty-one years earlier. Whatever his faults, Henry was a man of great resource and tenacity when conducting sieges, and the measure of his determination is shown by his refusal to leave Kenilworth and go to Windsor for the marriage of his daughter to the Duke of Brunswick; instead the Duke had to proceed to Kenilworth and receive his bride in the royal battle headquarters. The only fault that can be properly ascribed to Henry in this siege is that he did not begin it soon enough. Simon de Montfort the Younger arrived there on the night of August 1st, 1265. The decisive battle took place at Evesham three days later but the siege did not become total until the spring of 1266. The interim period had been occupied by Simon in introducing more supporters into the castle, and slipping out at intervals to pay morale-raising visits to other centres of resistance, even as far afield as the Cinque Ports.

The initial attack was launched with siege engines over the dry-ditch on the north side, the rest of the castle being considered virtually unassailable because of the width of the lake. The narrowness of this front cramped the efforts of the besiegers and it is said that the return fire from the castle frequently smashed the royal missiles in mid-air. As the stone projectiles were about eighteeen inches in diameter the odds against this occurring frequently were considerable but it undoubtedly did occur and not only at Kenilworth. The most improbable events do happen in warfare and all others fade into obscurity against the instance of the sniper's bullet which was met and stopped half way up the rifle by an enemy bullet in the First World War.

The failure of the artillery barrage over the dry ditch at Kenilworth was matched by an equally humiliating failure to cross it. Controversy over the relative merits of dry and wet ditches was as acrimonious and stubborn as over many subsequent military alternatives but, as far as Kenilworth was concerned, the advocates of the dry-ditch certainly had a good example. An attempt to mount a water-borne assault was completely disastrous, and the attackers were further harassed by sallies which destroyed royal siege engines and observation towers.

As the castle was still holding out in October 1266 an attempt was made to hasten its inevitable surrender by offering mild terms. These were rejected, and the King summoned the resources of the arsenal at Northampton to mount a final crushing assault on the weakened garrison. Before the final bloody encounter could take place a new factor in the form of disease raged through the castle garrison. It appears to have been typhoid and its result was that on December 12th Kenilworth surrendered, taking advantage of a clause in the original truce terms which had allowed forty days for second thoughts.

Although less spectacular occasions, the other pockets of resistance outlasted even Kenilworth but finally opposition was quelled by a general amnesty. Prince Edward departed on a Crusade, and the last five years of Henry III's long reign were the most peaceful of all.

Henry III had the distinction of keeping a white bear and an elephant in the Tower of London. The elephant was a present from France and the bear came from Norway. The bear was kept on an iron chain and muzzled but was allowed to go fishing sometimes when it would merely be attached to a rope. It and its keeper had a ration allowance of 4d. a day. Usually monarchs preferred more regal animals: Henry I had lions; Henry II had leopards; Edward I and Edward II were contented with lions; but Edward III had a lion, a lioness, a leopard, and two tigers.

* 9 *

The Great Era of Castle-Building

Edward I (1272–1307)

AT the time of his accession Edward was on his Crusade, but his abilities were already well known and the fact that he did not reach England till nearly two years later made no difference to the stability of the government carried on in his name. The confidence placed in him was fully justified; privately and publicly he was a model of restraint and good conduct, and his reign was marked by justice, efficiency, and generosity to everyone except Jews.

Edward I's contribution to the art of war was considerable, and the effects long-lasting. He was the first to appreciate the merits of the longbow and his development of this weapon laid the foundation of success that came long after his own. The longbow spelt the doom of the knight in heavy armour; after Edward there was no such thing as immunity on the battlefield.

At the outset he faced considerable tasks. The Statute of Mortmain was designed to prevent the evasion of taxation under the connivance of the church, and of Quo Warranto to recover lost royal lands. The latter statute, which sent a commission to enquire 'by what right' land was held, aroused much opposition among the barons. Notable among the opposers was John de Warenne, Earl of Surrey, who dramatically brandished a rusty sword, claiming, probably rightly, that the sword of his ancestors had won his lands, and by the sword of his ancestors he would hold them.

In general, Edward tightened the feudal system by linking all ranks more closely together. He accomplished this by changes in the land laws. By the Statute of Winchester in 1285 he reorganized the national militia, laying down what arms each man should furnish for himself. The same act established a 'watch', or local police, to control robbers.

But the main preoccupation of any mediaeval king was to wage successful war. For Edward this fell into three separate phases: Wales, Scotland, and France. Of these Wales took prior place.

By the thirteenth century the southern and border districts of Wales were under English control. These areas were known as the Marches, and their controllers as the Marcher Barons. But in the north and west it was a very different story. The Welsh princes stayed in their inaccessible territories when the English political situation was stable, but at the first signs of trouble over the English border there would be deep and savage raids. Fortunately for England the Welsh princes were occupied by their own bitter feuds; and when an English king invaded Wales he could usually count on the assistance of at least one prince who hated his neighbours more than he hated the invader. A similar situation occurred in Ireland.

However, the Welsh were bound to do homage to the English king, and when Edward ascended the throne he summoned Llewellyn, ruler of North Wales, to his court for that purpose. Llewellyn had too long a record of success against the English to take kindly to this invitation. He had profited considerably from his alliance with Simon de Montfort, when he had made numerous border raids, and in 1275 made an unsuccessful attempt to marry Simon's daughter; the plan miscarried because the ship bringing her from France was captured by Edward.

This last gesture of defiance was too much for Edward who decided that military measures were the only solution. The news was welcome to the Marcher Barons, whose chain of border castles was all too vulnerable to Welsh attacks, as they knew from experience.

The invasion, when it came, was methodical and powerful. The castles of Flint and Rhudlan were repaired and strengthened, and from these secure points Edward drove cautiously but steadily into North Wales. The natural defence of inaccessibility was overcome by building roads into the most remote areas. By this means Edward was able to establish a strangling blockade on the hill fastnesses, a siege on a vaster scale than had ever before been visualized, and which anyone but Edward might have thought impossible. After several months the Welsh were facing starvation, and Llewellyn surrendered. He was granted more generous terms than he expected, was allowed to marry De Montfort's daughter, and was reconciled to his brothers, whom he had driven out of his principality. As a result David, the eldest, was granted a barony by Edward. The first round in this long contest ended in 1277.

But old habits die hard. Five years later, with supreme ingratitude, David joined Llewellyn in a fresh undeclared war. Between them they ravaged the northern coast as far as Chester. Edward was taken by surprise, but not for long. Once more he assembled a vast army, and once more drove Llewellyn back into the remote fastness of Snowdon. Even then he was not ungenerous, for he offered Llewellyn his life and an English earldom. But Llewellyn staked all on a last desperate throw. With a select band he slipped through the English lines at night and tried to assemble a force in central Wales. Here luck deserted him, for he was killed by an English squire in a minor skirmish. David continued to hold out but starvation soon drove him to capitulation. He was tried as a traitor, for when driven out of Wales by Llewellyn he had lived in the English court. He was executed barbarously (he was pulled apart by horses) and his head displayed beside his brother's at the Tower.

The end of the first phase of the Welsh wars in 1277 was also the beginning of a vast programme of castle building and strengthening. It was accompanied by a less spectacular but equally thorough project of road-building. The overall strategy was to secure the invaders' communications by a chain of roads and castles across the lifeline of any potential Welsh resistance. The import of food, weapons, or allies was thereby checked, and the interior mountains, instead of being a springboard for attack, would become a dangerous trap where starvation would achieve more than the sword. In that year Edward initiated the complete reconstruction of Flint, Rhuddlan, Aberystwyth and Builth. The first three of these were sited so that they could be supplied from the sea, although in the case of Rhuddlan this meant diverting the river Clwyd, a task of great magnitude and expense. Although Rhuddlan is now less well known than such castles as Conway and Harlech there is no doubt of the importance which Edward I attached to it. It was designed to replace Chester, and its architect, James of St George, also supervised the construction of Conway, Caernavon, Harlech, and Beaumaris. The new castle stood a little apart from the old motte and bailey structure, and was of new concentric design. A concentric castle broke away from the old pattern of linear defence through barbican, gatehouse, outer ward, and inner ward to keep, and instead had a system of enclosed squares. The inner fortifications overlooked the ring immediately outside them and it would therefore be possible to concentrate maximum fire power on any point of attack.

Beaumaris is the finest example of the Edwardian concentric castle, although Caerphilly, which was begun by the Earl of Gloucester in 1265, and completed by Gilbert de Clare in 1272, is certainly its equal. The central court at Caerphilly was fifty feet above the outer ward, and it is not surprising that it was never taken by force. Of Aberystwyth and Builth little now remains to show what they once were but it is unlikely that they were much inferior to their contemporaries in strength. Builth may well have been the strongest of all for it took four years to construct, was designed by a notable architect, and though frequently besieged never surrendered—no mean feat for a castle in the middle of Wales.

When the first four castles were nearing completion, Edward commissioned the next three, which were at Conway, Caernavon, and Harlech. The earlier castles would not have cost less than a million pounds at present-day values, but this second ring, to which Beaumaris would be added in 1295, would cost at least £3 million—perhaps twice as much. They were not a mere local enterprise. Not only England but also Europe was scoured to obtain the right quality and quantity of designers and masons. Each castle probably absorbed 10 per cent of the national work force so the economic strain that this programme imposed may easily be imagined.

Although concentric defence was the ideal, it could not always be used owing to the nature of the ground. Caernavon and Conway were not concentric owing to the lie of their sites; however they were in their own way as formidable as Beaumaris which was built on classic concentric lines. Caernavon is said to be the most palatial of Edward's castles, Conway the most magnificent, Beaumaris the most efficient, and Harlech the most beautiful in its setting.

Needless to say the Welsh did not view the rise of these symbols of their subjugation with any pleasure, and took appropriate action. In 1294 Prince Madog lynched the Sheriff of Caernavon, and burnt a portion of the new buildings. Denbigh, which has a most impressive gatehouse, was captured the same year, but Flint and Rhuddlan, though attacked, were able to sit out their sieges without difficulty. The rising was soon suppressed.

The real test of Edward's castles did not come until nearly a century after his death. In September 1400 Rhuddlan once more came under siege but withstood it. Harlech was not so fortunate. It was captured in 1405 and became Owen Glendower's capital for three years. Caernavon withstood sieges in 1403 and 1404 although its garrison was down to 28 men. As the attackers lost 300, the reputation of Caernavon was seen to be justified.

Among the many impressive features of Edwardian castles was the architectural one of the 'battered' tower. This does not mean, as well it might, the effects of a siege engine but describes the very slight taper which was given from the base to midway, and then imperceptibly moved to the perpendicular. It added considerably to the strength, but the use of the term 'battered' has proved remarkably confusing (e.g. battered plinth).

While Edward built his castles to make all parts of Wales accessible, the Welsh built theirs to produce the opposite effect. The Welsh were not greatly concerned about being starved into surrender in their own strongholds for they considered that any besieger would be in a desperate plight if he tried to live off the surrounding countryside. One of the more daunting Welsh castles to-day is Dinas Bran. Overlooking the town of Llangollen it also dominates the surrounding valleys. It is only accessible on one side and that used to be protected by a deep ditch. Anyone who crawls to the ancient ruins on top might well wonder how such a place could ever be taken, but taken it was in 1282, and given to Earl Warenne of Surrey, who must have found it a change from the benign surroundings of Lewes or Reigate.

Three remarkable Welsh castles are Dryslwyn, Caer Cennan,

and Dynevor. Drysllwyn and Dynevor are five miles apart, each on a steep hill overhanging the north bank of the Towy. Caer Cennan is five miles south-east of Dynevor. Dynevor, although lowest of the three, was not for that reason any more accessible; on three sides it had precipices and the fourth was a steep slope with dry ditches at intervals. All are within a few miles of Carmarthen but can scarcely be said to be within easy reach of it.

Drysllwyn was the scene of a remarkable siege in June 1287. Its owner, Rhys of Drysllwyn, decided to rebel against Edward. There was no apparent reason for this decision, for Rhys had previously served the English king loyally and helped him subdue various rebel Welsh leaders. The task of disciplining Rhys fell on Edmund of Cornwall, who was acting as Regent while Edward was in France.

Edmund surrounded Drysllwyn Castle with 1100 men, but his greatest asset was a huge siege engine. It appears to have been a trebuchet, and was escorted by 20 horsemen and over 450 foot soldiers. It was conveyed on four-wheeled wains dragged by forty oxen when the ground was easy, but required sixty for difficult terrain. The ammunition of 480 stones was carried by a packhorse train. In the twenty-day long siege it was in constant use, and was supported by the activities of miners. Some of the latter were too ardent and brought down a portion of wall prematurely, killing a number of their own knights who were caught unawares when it fell.

Rhys escaped from Drysllwyn, and in November captured Emlyn castle by surprise. Once more the massive siege train set on its way, but it was mid-January and the worst of winter before Emlyn fell. Rhys had also planned to make use of Dynevor and Caer Cennan, but abandoned this idea for reasons unknown. Caer Cennan looks impregnable but in practice was not so. Like many Welsh castles it had precipices on three sides and a near precipice on the fourth. It had first come into prominence in the period of Llewellyn. In 1277 a Marcher Baron, Pain de Chaworth, had recaptured it. After 1282 it was dismantled and destroyed, only to be repaired soon after and garrisoned with 500 men by the Earl of Gloucester. As noted, Rhys regained it for the Welsh but not for long. For a hundred years it was unrecorded, but when Glendower held the rest of

Carmarthenshire it stood a siege of over a year under a gallant but almost unknown castellan called John Skidmore. Subsequently it fell into ruin, became a haunt of outlaws, and was ultimately made uninhabitable by the Carmarthen sheriff in the mid-fourteenth century.

Scotland was the second of the territories on which Edward had claims. When Edward became king, Alexander III of Scotland attended his coronation and acknowledged that he was the liegeman of his English lord. This was an act of homage rather than an acceptance of English feudal superiority. In 1286 Alexander was carried over the cliffs of Kinghorn by an unruly horse, leaving as his only surviving descendant a granddaughter of four. This little girl was in Norway at the time and for three years a Regency ruled in her name. It was decided in 1290 that she should be brought to Scotland and betrothed to Edward's son, the Prince of Wales. Accordingly she embarked for Scotland in the autumn of the same year. Tragically, the ship was tossed around for weeks in North Sea storms, and when at last the princess landed in the Orkneys, she died from the effects. With the death of 'the Maid of Norway' all hopes of a stable relationship between England and Scotland disappeared for many years. In Scotland the new rival claimants to the throne, John Balliol and Robert Bruce, had equally sound claims for consideration. The Scottish barons asked Edward to appoint a court of arbitration, and this body approved John Balliol as the stronger candidate.

Unfortunately the character of Balliol was not as strong as his legal claim, a point which had been duly noted by Edward, who applied increasing pressure on him to bring Scotland under English control. When the Scots, including Balliol, found this subservience intolerable they negotiated an alliance with the French, who also had their grievances against England. Assuming that Edward would be too occupied by his French problems —and the Welsh who were also up in arms—Balliol collected an army of some forty thousand to invade northern England. After some preliminary successes this force was checked at Carlisle. By this time Edward had, as usual, reacted swiftly to this dangerous situation and had a considerable army on the road to Berwick.

The town was well garrisoned by picked Scottish troops, and

the flower of these held the castle, but both were overwhelmed by the desperate onslaught that assailed the town. Attack by sea resulted in heavy English casualties and had little but a diversionary effect; the land battle was the decisive factor. Edward was in the forefront, and was first into the town. The subsequent massacre lasted for two days, and was horrifying even for those hardened times.

But if Edward thought this vigorous onslaught would drive the Scots into submission he was soon shown to be seriously mistaken. Balliol sent a message that amounted to open defiance.

The next step was to recover Dunbar castle, and the task was given to John, Earl of Warenne, alluded to earlier in connection with Quo Warranto and the rusty sword. Here was a curious situation, for the castle had been handed over to the Scots by the Countess of March although her husband was fighting in Edward's army. When the Scottish army advanced to its relief Warenne intercepted and routed it. It may seem inconceivable that the main Scottish army could be routed by a mere English vanguard, but the fact that it happened shows how little confidence Balliol inspired. After the principal action the castle had no alternative to starvation but to surrender.

From then on the war became a matter of reducing Scottish strongpoints, a process that, now the main army was broken, took only two months. But Edward's calculation that the war was over was again seriously at fault. As Warenne settled down to his task as governor, and Cressingham, the Treasurer, and Ormesby the Justiciar to theirs, a fresh insurrection was already brewing.

The leader of this much more serious military movement was William Wallace, who was perhaps a greater leader and fighter than Edward himself. Under happier circumstances his qualities could have brought much benefit to both countries, but at the time they were wholeheartedly devoted to hatred of the English yoke and all that went with it. It was not long before other notable fighters rallied to his cause, many of them breaking oaths of fealty to Edward to do so. Among them was 22-year-old Robert Bruce, grandson of the unsuccessful claimant to the Scottish throne. Wallace's force was soon able to retake many lost castles, and was actively besieging Dundee when the English army approached Stirling.

The Scottish victory that ensued is blamed entirely on Warenne who, over-confidently, had allowed his army to cross the Forth by a narrow wooden bridge. After about half the force had crossed, Wallace's men swept down on them with a vigour that would have been hard to withstand in the open field; in the unfortunate position of the English it was unstoppable. Victory was complete; all waverers joined the Scottish cause, and the remaining strongpoints in Scotland surrendered. Scotland was lost.

By the spring of 1298 Edward, although unsatisfied in his French aspirations, returned to England. He lost no time in assembling an immense army at York, and soon moved northward. But surprisingly, as he advanced, no Scots were seen. Even when he reached Roxburgh there was still no enemy and, more serious, no sign of the supplies he had sent in advance to Berwick. At this critical point came news that the Scottish army was encamped at Falkirk. In spite of two broken ribs caused by a kick by a horse when he was sleeping on the ground, Edward led his army forward. The English had a considerable advantage in numbers as well as in variety of arms and inflicted an overwhelming defeat on the Scots. It was a barren victory. Retreating Scots burned everything behind them with the result that Edward was shortly forced to retire because the countryside afforded nothing to victual an army. The Scottish barons deposed Wallace and elected John Comyn (a relation of Balliol) to succeed him. Edward turned his attention to France again. First, however, he had to meet and and settle considerable criticism over the cost, and hence taxes, of these foreign wars. Even so the murmurings of his discontented subjects were only temporarily quietened when he set off: they were to be renewed with greater force when his weak son succeeded him. However, the French impasse was resolved by marriage and not by war. The Pope, as mediator, proposed that Edward, long a widower, should be married to the French King's sister, and that Edward, Prince of Wales, now aged 13, should be betrothed to Isabella, daughter of the French monarch: the latter proved a disastrous match. These diplomatic arrangements were completed in September 1299.

Two months later Edward set out on another march to Scotland, a march that was almost immediately halted and turned into an ignominious withdrawal without a blow being struck. His barons had refused to support him in a policy so costly that it implied a royal dictatorship as far as finance was concerned. The principal effect of this was that the garrison of Stirling Castle gave up hope of relief, and surrendered to the besieging Scots. During the next three years Edward made an annual invasion with moderate but inconclusive success; on the fourth he sent a deputy with twenty thousand men, which resulted in complete defeat and disaster. This was February 1303.

By the summer of 1303 Edward was able to withdraw his forces from France and give full attention to his Scottish ambitions. He drove a path of destruction from Edinburgh, Perth, and Aberdeen, to Kinloss. At this stage he established an impregnable headquarters at Lochendorh, an island in the middle of a lake, where he accepted a general surrender of the north. This accomplished, he moved south meeting no resistance other than at Brechin.

But Stirling, the centre and hope of Scottish resistance still held out. The ensuing siege lasted from April 22nd to July 20th, 1304, and as the castle was defended by a mere 140 men must rank as one of the most desperate defences in history. Edward, the expert in castle defence, was equally resourceful in attack, and spared no pains to bring overwhelming force against the castle. Thirteen siege engines were constructed, some from material removed from the nearby St Andrews cathedral. The fact that they could project stones weighing up to three hundred-weight gives some idea of their power and the force of the assault. Edward was as prominent, though more fortunate, than Richard I had been, and was wounded several times. His wife watched the siege from a specially constructed balcony.

As the first assaults were repelled Edward sent south for fresh weapons from as far back as London. With this increased armoury the besiegers were soon attacking what was little more than a pile of rubble. Very wisely Edward was magnanimous in victory, and the lives of the surrendered garrison, by this time reduced to 30, were spared. Unfortunately this clemency was not extended to Wallace, who, shortly afterwards, was captured. His execution at Smithfield was carried out with every humiliation and brutality that barbarism could conceive.

But Edward's wars with Scotland were not quite over. Resistance again flared up, and this time the leader was Bruce. Edward, now 69 years old, was too ill to take the lead, and travelled behind on a litter. He was suffering from a mortal illness but directed operations from his sick bed until the frustrations of a long-drawn campaign drove him once more to horseback. But it was the last time. He could make slow progress only and within a week was dead, to his last breath spurring on his followers to complete the conquest to which he had devoted most of his reign, time, and country's resources.

✻ 10 ✻

Favourites and Foreign Wars

Edward II (1307–1327): Edward III (1327–1377)

EDWARD II managed to combine the worst qualities of the Plantagenets without possessing any of their virtues. There is no apparent explanation for Edward. His father was, for his age, an exemplary king and man. His mother, Eleanor of Castile, was a woman of great virtue, who is commemorated on the English scene by the crosses which marked the resting places of her coffin on the journey from Grantham to Westminster; the best known is Charing Cross. Perhaps the fact that she died when he was only seven may have contributed to the fact that he grew up idle, obstinate, feckless, and at times appallingly cruel. Yet the picture is not entirely dark; he could be extremely courageous, and also kind and thoughtful. Disastrously for him, and for the country, he had a passionate attachment to the most unworthy favourites. The homosexual strain that had first appeared in William Rufus, and reappeared with Richard I, was now going to lead the reigning monarch straight to disaster.

The object of his affections was Gaveston, an adventurer from Gascony. Edward did not stop at lavishing affection on him; once in power he also conferred titles such as Earl of Cornwall, and appointments such as Lord Deputy of Ireland.

Gaveston thoroughly enjoyed his position, and the honours which the doting king bestowed on him. He was hated by the barons partly because he was an upstart, but more because he returned their hatred with contempt. He tore them to pieces with his tongue, and when they took part in a tournament, defeated and humiliated four of them in the lists.

The tournament in question took place at Wallingford Castle, and was held outside the walls, for it was never considered advisable to allow large numbers of armed knights inside the wards; elsewhere, on previous occasions, in temper or treachery knights tried to take possession of the host castle. The Wallingford tournament was held to proclaim the glory of Piers Gaveston, and the senior barons therefore decided it would be an excellent opportunity to make it a shameful occasion for him. Unfortunately for them Piers and his supporters were more youthful and agile, and won all the events. However, mediaeval custom decreed that the side which lost most times and was unhorsed most frequently was braver and stronger and was therefore the winner. (*Nam ipsius ludi lex esse dinoscitur quod qui plus perdit et qui saepius ab equo deicitur probior et fortior judicatur.*) This extraordinary law was doubtless designed to encourage people to remount till they could remount no more, and would therefore develop fortitude for battle. It may also have laid the foundation for that extraordinary English attitude that it is better to be a good loser than a good winner, that Dunkirk, Crete, and Singapore were in some mysterious way better than victories, and that the only creditable success comes effortlessly. Such a philosophy is appropriate if a person or nation wins most of the time but looks slightly tarnished if defeat is too frequent.

Edward's infatuation went beyond the bounds of common prudence. When he went to France to bring back his bride, Isabella, Gaveston was left as Regent; at the wedding in Westminster Gaveston took precedence over all others; when the nobles demanded that Gaveston should be banished Edward complied but made him Lord Deputy of Ireland. Finally, the point was reached when the entire country was solidly united in hatred of Gaveston, and the king was compelled to exile him. But the banishment was shortlived, and the barons decided to take matters into their own hands. Taking advantage of the temporary separation of the King from his favourite they besieged the latter at Scarborough. Foolishly, for he was in a strong position, he surrendered under a pledge of safe conduct. On the way to Westminster he was betrayed to his great enemy the Earl of Warwick, 'the black dog of Arden', who executed him summarily on Blacklow Hill.

While Edward was undermining his own position in England, Bruce was making great headway in Scotland. After the siege of Perth in 1312 Stirling was the only fortress held by the

English in Scotland which was capable of putting up a lengthy resistance; before long Bruce was over the border and ravaging as far south as Chester. Although primitive and ill-equipped, the Scottish forces were endowed with great courage and endurance, while clever leadership compensated for their lack of military resources. The castles of Roxburgh, Edinburgh, and Linlithgow were all won by stratagem and not by open assault. At Linlithgow soldiers were smuggled into the castle under a load of hay and the cart used to jam the gates. By this stage it looked as if, instead of the English subduing Scotland, the Scots were more likely to conquer England. The moment of national danger meant that this was no time for settling personal differences, and the killers of Gaveston were pardoned, though not perhaps forgiven.

The full machinery of the feudal 'call-up' was used for the crisis army. Urgency was given by the fact that the Governor of Stirling, Philip de Mowbray, had agreed to surrender on June 24th if not relieved before. A well balanced army, whose exact numbers are not known but which nobody quotes as less than 100,000, assembled at Berwick.

Edward then marched for Stirling by a route which Bruce was able to forecast accurately, and therefore prepare. But even Bruce could hardly have envisaged the stupidity with which Edward would handle his troops, who were caught in a bad position between concealed pits and a morass at Bannockburn. Archers and heavy cavalry were completely wasted; the only point to Edward's credit is that in the final general rout he refused to fly till he was dragged from the field by the Earl of Pembroke. A curious point of chivalry arose when he attempted to take refuge in Stirling Castle; the governor informed him that he had given his word to deliver the castle if not relieved by June 24th. Edward rode on.

The English lost at least 10,000 men and a mass of valuable equipment and provisions; the Scottish losses were about four thousand. It was a disastrous day for England and, as it turned out, for Scotland. In spite of the continuous fighting England and Scotland had been drawing closer together, and were near a form of unity and friendship, but Bannockburn destroyed all hope of this for three centuries. Victory made Scotland a nation and the result was endless border warfare, plunder, and misery, until in 1603 James VI of Scotland also became James I of England.

Bannockburn was not the only calamity England sustained in 1314, for the same year saw a bad harvest, and the next year a worse one. Cattle disease and pestilence added to the general misery. Berwick was lost to Bruce in 1318, and an attempt to recapture it failed. The Scots crossed over to Ireland with the intention of ousting the English but failed to do more than confine them to the eastern coast—the Pale. Eventually a truce was signed in 1320.

But freedom from immediate external danger merely gave an opportunity for internal discontents to flare. These centred on an enormously wealthy family—the Despensers, who were made even richer by Edward's presents. It was alleged that they incited Edward to foolish, corrupt, and illegal acts, although he could hardly be said to need encouragement, and in 1321 Parliament drove them into exile.

On this occasion Edward showed considerable cunning. His first move was to collect an army with the proclaimed intention of punishing Baron Baddlesmere, who had refused Queen Isabella entry into Leeds Castle, in Kent. The pretext was an incident that occurred in the middle of 1321. Queen Isabella had appeared at night before the castle gates with a large retinue and demanded entry. The Constable, one Walter Colepeper, refused somewhat brusquely, saying she could not enter without orders from Baron Baddlesmere. The Queen ordered an immediate attack but it failed. Edward II proclaimed a levy from the neighbouring four counties, and on October 23rd the siege began. Baddlesmere tried to relieve it, first by negotiation, and then by a diversion staged at Kingston, but both efforts failed. Aymer de Valence, Edward's commander, prosecuted the siege with such vigour that the castle was forced to surrender on the eighth day. Colepeper and twelve others were hanged, Lady Baddlesmere put in the Tower of London, and the baron, when eventually captured, was executed. Lesser ranks were merely imprisoned.

The loyal but luckless Colepeper would have been glad to know that 300 years later his family would purchase the castle, be created barons, and, although dying out in the male line, be associated with it for a further 300 years.

(*Mansell Collection*)

1. Siege Warfare as seen by a mediaeval artist—the siege of Duras in Froissart's Chronicles. Note the pavas protecting the crossbowmen and the scaling party.

(*Mansell Collection*)

2. Siege Warfare: the Belfry. A belfry was useful (i) as a watch tower, (ii) as a vantage point for aiming incendiary arrows and (iii) for assaulting the battlements. A further advantage was that, while the defenders of the battlements were strongly engaged, the battering-ram party below was able to work virtually undisturbed.

(*Mansell Collection*)

3. Siege Warfare. Note the mine and countermine. In front of the walls the ground has been excavated and the roof—shored up with props—is ready to collapse under the weight of the belfry when it is moved forward.

(*Mansell Collection*)

4. Siege Warfare: the Crow. This ingenious device could swoop in the flash of an eye, and was used for hooking up besiegers who became careless while on reconnaissance. Such prisoners were often valuable as hostages or, under torture, as sources of information.

In 1139 Prince Henry of Scotland narrowly escaped being hooked up at the siege of Ludlow. See page 97.

(*Mansell Collection*)

5. Siege Warfare. An early form of parachute drop! The advantage of this siege engine lay in its capacity to place a task force behind the defences at great speed.

(*Aerofilms Ltd.*)

6. Portchester Castle, Hampshire. The Roman fort was constructed in the third or fourth century. After 1160 Henry II reconditioned the Roman walls and built the square Norman keep, originally measuring 65 ft. by 52 ft. and 100 ft. high.

(Aerofilms Ltd.)

7. Rochester Castle, Kent, was the scene of several notable and unusual sieges. Note the circular design of the south-west tower, rebuilt in this form after the earlier square tower of Bishop Odo's keep had been destroyed by siege operations.

(*Aerofilms Ltd.*)

8. Dover Castle, Kent, has commanded the sea-crossing from the Continent for over 2,000 years. The Norman keep and bailey can be seen inland from the earlier fortifications which enclose a Roman lighthouse and church of Saxon origin.

(*Mansell Collection*)

9. Dover Castle. This gatehouse shows modifications made during the fourteenth century.

(*Aerofilms Ltd.*)

10. Restormel Castle, Cornwall. Built about 1110. The site, which commands the River Fowey, was probably chosen by Robert of Mortain. The shell keep is 110 ft. in diameter and designed with all the windows of the lower storey facing inwards. The donjon has now disappeared. The castle was altered during the thirteenth century. The projection on the right of the photograph is a gatehouse; on the left is a chapel.

(*Aerofilms Ltd.*)

11. Caerphilly Castle, Glamorgan. Begun 1267, overthrown by Llewellyn, and completed by Gilbert de Clare by 1275. As at Kenilworth, two streams were dammed to form the moat. The inner ward of the castle rises 50 ft. above the outer. Note the strong barbican and tower protecting the sluice gate. Behind the main fortifications on the island is the 'hornwork'—an extra outer bailey of which little remains.

(*Aerofilms Ltd.*)

12. Chirk Castle, Denbighshire. Note the massive drum towers of this Welsh Edwardian castle.

(*Aerofilms Ltd.*)

13. Harlech Castle, Merioneth. This was probably the most formidable of the Edwardian castles, standing high above marshland and water. Not surprisingly it sustained the longest siege in British history.

(*Aerofilms Ltd.*)

14. Beaumaris Castle, Anglesey. Built between 1295 and 1323, this fortress is militarily the most sophisticated of the Edwardian castles. Note the concentric plan and the narrow confines of the outer bailey, completely dominated by the inner walls.

(*Aerofilms Ltd.*)

15. Warwick Castle. Note the formidable gatehouse and the tall flanking towers, 128 ft. and 147 ft. high, built in the late fourteenth century.

(*Mansell Collection*)

16. Herstmonceux Castle, Sussex. More of a gracious country house than a grim fortress, it was built in the latter part of the fifteenth century. The grounds now provide the home of the Royal Observatory.

It is somewhat of a mystery how an immensely strong castle like Leeds can have been taken in so short a time. It was surrounded by fifteen acres of water and had a strong gatehouse protected by two barbicans. The probable explanation is that the garrison was too small for such a wide circumference. Aymer de Valence was Earl of Pembroke but had no kinship with the Earl of Pembroke who had captured the castle in 1138. In its earlier days the place was known as Ledes or Escledas; it has no connection with Leeds in Yorkshire although both names are derived from the Anglo-Saxon for 'stream'.

When the garrison at Leeds had been disposed of, Edward turned on Thomas, Earl of Lancaster, who had been one of the chief opponents of Gaveston and had recently made an alliance with Bruce. The expected help from Scotland never came. Consequently Edward was able to seize his powerful enemy at Pontefract Castle and have him executed in public just outside the town. Another forty of Edward's opponents were executed also, and the royal power was supreme again.

A year later there occurred a curious incident which demonstrated the change that had taken place in the relationship of the castle to the local community. Although most of the King's leading opponents had been executed or put in prison in 1322, others remained who were ready and willing to take their places. A group of them made a plan to seize certain royal castles in 1323. On the list was Wallingford, where Maurice de Berkeley was imprisoned. Maurice was allowed a number of privileges, one being a fairly free choice in the number and type of his visitors. One evening he entertained a few companions and also invited in the Constable, the watchmen, and the doorkeepers. Halfway through the meal Maurice's cronies demanded that the Constable should give up the castle keys; the Constable, caught unawares and unarmed, had to comply. The conspirators then let in a further twenty of their supporters but at this point their plans went wrong because a boy decided that something was amiss and slipped away quietly to the Mayor of Wallingford. The Mayor lost no time, bells were rung, horns blown, and the castle surrounded. Within hours the Sheriff was demanding the surrender of the castle. At first the conspirators tried to bluff him by pretending they were executing a royal command but when he asked for proof they gave in and admitted his forces.

But the King had learnt nothing from his setbacks or triumphs. He abandoned the administration of the kingdom to the Despensers, a father and son, both called Hugh. Haughty, stupid, well-hated in their own right, they contributed greatly to the unpopularity of the King.

But unpopularity alone would not have dethroned Edward. The conspiracy that did so occasions some sympathy for him, although in view of his male favourites it was hardly surprising. His wife Isabella was in France at her brother's court on a diplomatic mission, when she met Roger Mortimer, an exiled Marcher baron, and became completely infatuated with him. With Mortimer and a host of Flemish mercenaries she returned to England with the intention of deposing Edward and ruling through her son. Her army was soon joined by disaffected barons and their forces. The elder Despenser was captured at Bristol. He was aged ninety but received no mercy, and his life came to a dramatic and unpleasant end. Instead of being hung, drawn and quartered, he was drawn, quartered, and then hung; the pieces of his body were chopped up and fed to dogs after four days. The younger Despenser had an equally spectacular execution, though it differed in some details.

Edward tried to escape by sea but failed. He was eventually taken to Berkeley Castle in Gloucestershire where attempts were made to cause him to die from disease—the castle sewage ran through his cell. But his constitution was too strong for disease to work quickly enough, and he was murdered one night and buried soon afterwards. The room in which the crime occurred can be seen to-day, for the castle has been lived in, and not allowed to fall into ruins.

Edward III was only 14 when, nominally, he became King of England. His father's fate was not widely realized, and the people had little suspicion of the true state of affairs between Isabella and Mortimer. The chain of command was that the young King was ruled by the Earl of Lancaster, Lancaster by Isabella, and Isabella by Mortimer. This arrangement worked reasonably satisfactorily until the renewal of the Scottish war. A thirteen-year truce had been signed in 1323 but the weakness of England was more than Bruce could stand idly by and watch. The fact that he was old and a leper made no difference to his resolve. In 1328 he sent James, the 'Black Douglas', with some 25,000 horsemen on a border raid that devastated Northumberland and Cumberland, but he was never brought to battle. Mortimer was so out-generalled and out-marched that he persuaded the Queen to make peace with the Scots, the terms of which were so humiliating that the treaty came to be known as the 'Shameful Peace'. The young king accompanied Mortimer on this lamentable campaign and must have been impressed by his utter incompetence.

Edward was growing up. At the age of fifteen he had been married to Philippa of Hainault, and two years later she had borne him a son, the future Black Prince. As a King and a father he found his position intolerable. Plainly the first step must be to get rid of Mortimer and with this in mind he carefully prepared a scheme for his arrest in the autumn of 1330. The court was at Nottingham Castle, and the keys were taken to Isabella every night: but there was a secret subterranean passage leading from the west side of the castle rock, and through this one October night came a powerful force led by the Earl of Montacute. The invaders joined by young Edward, went to the Queen's apartments, and captured Mortimer after a brisk skirmish. Isabella pleaded and cursed, but to no avail. Mortimer was executed soon afterwards but the Queen, after a plea from the Pope, spent twenty-eight years in confinement at Castle Rising.

During the first part of his reign Edward III was extremely popular and hopes ran high. Unfortunately he had the same warlike ambitions as his grandfather without the latter's statesmanship and, in his last years, he fell into premature senile decay, surrounding himself with favourites as obnoxious as those who had caused the unpopularity of Edward II.

His opening move was to revenge Bannockburn, and the confused situation in Scotland soon gave him a chance to do so as the champion of the deposed Balliol. At Halidon Hill, near Berwick, in 1333, the English archers were able to slaughter the Scottish pikemen and infantry with little loss to themselves, for on this occasion there was marshy ground between the two armies.

But Edward's main ambitions lay in France, and in 1337 he claimed the French crown on quite untenable grounds. His opening campaign in France was ineffective but he made some headway when the French collected a fleet for a counter invasion. Edward caught them in harbour at Sluys, induced them to put to sea, and then defeated them by skilful archery, and boarding. But it was an isolated triumph, for his next invasion of France achieved as little as the first.

In 1341, while Edward was besieging Tournai, the Scots accomplished a memorable feat in retaking Edinburgh Castle. Sir William Douglas and about a dozen of his companions disguised themselves as peasants and approached the castle with a dozen pack-horses, carrying wheat, oats, and coal. Nearby in the abbey ruins 'two hundred of the wild Scots were hidden'. It was very early in the morning and the porter hesitated to wake the steward but instead decided to let the travelling salesmen wait in the outer ward. As soon as they crossed the threshold they dumped the coal in the gateway so that it could not be closed 'and then they took the porter and slew him so peaceably he never spake word'. Then Douglas blew his horn and rallied the rest of his supporters. Before the obstructions could be cleared and the gate closed reinforcements were in the castle, and the garrison was soon slaughtered.

Froissart gives much interesting detail of the siege of Reole in 1345. Two large belfries, each moving on four wheels, approached the walls. Each belfry had three stages and each stage contained 100 archers who fired devastating volleys in complete unison. The object was to protect the two hundred miners who hacked a way into the wall that lay between the belfries. These tactics were successful but the siege was more arduous than had been anticipated.

At Aiguillon in 1346 a reward of a hundred crowns was offered to the first to reach the gatehouse. The assault troops responded to this incentive so briskly that they threw each other into the river in order to be the first, and thus the winner. The fact that the defence was showering the area with quicklime, boiling water, and lumps of iron does not appear to have interested the competitors.

In 1346 Edward decided to fight without continental allies, and set off with a force of about 20,000, of which half were archers and one-quarter mixed Welsh and Irish infantry. For several months it did nothing more constructive than lay waste the countryside, and arouse the fury of the French. The latter mustered some 70,000 and in August 1346 had caught up with Edward in Ponthieu. In the circumstances Edward decided that in spite of the disparity of numbers his best chance was to fight it out then and there.

The ensuing battle of Crecy was as astonishing to the victors as to the vanquished; about half the French force was killed and, although ineptitude by the French command certainly contributed to this, the greatest factor was undoubtedly the marksmanship of the English bowmen. Speed of fire was the decisive factor, and war took on a new pattern. The arrow took no account of rank or birth, and the French nobles fell as swiftly and finally as the common soldiers. The old order in which knights and nobles had enjoyed the fun without the danger was over for ever; henceforth equal risks would be taken by all.

Crecy is often mistakenly thought to have sounded the knell of the crossbow but this is widely inaccurate. It is true that the Genoese crossbowmen failed at Crecy because a heavy shower had wet their bowstrings, when the English longbowmen had managed to keep theirs dry. But under normal conditions the crossbow still had a much longer range and could be fired without special training. The failure of the Genoese archers infuriated the French knights, who rode in among them and slaughtered them wholesale, immediately prior to being slaughtered themselves by the longbowmen. This was not the only occasion in mediaeval warfare when impatient or frustrated knights cut down their own foot soldiers in order to clear a path to the enemy.

Edward was triumphant. Although he could not obtain the crown of France as a legal right he had pressed a strong military claim. All he needed now was Calais, a seaport which would give him entry to France whenever he wished. Accordingly, after the battle, he marched directly on the town.

But Calais was strong, and could not be taken by frontal assault; certainly not with the limited numbers Edward had to throw against it. Accordingly he decided to save lives and win the town by famine. On the land side he built a second town of wooden houses to accommodate his troops, while on the other his fleet blockaded the harbour and cut off the town's communication with the sea.

The French commander, John de Vienne, was as ruthless as his besieger, and his first step was to rid the town of non-combatants who would consume valuable supplies. Accordingly he thrust out of the town all those who were of no use to him. Edward received them chivalrously, gave them a meal, and money, and let them through. But when later a further 500 were pushed out he was not so chivalrous, and the wretched outcasts died of appalling hardships in the no man's land between the two forces. It was the story of Château Gaillard over again. The siege lasted a year, after which Calais became an English town for the next 200 years.

Long before the siege of Calais was over Edward had a further success over the Scots. In October 1346 the English archers once more slaughtered the Scottish pikemen, this time at Neville's Cross, north of Durham.

But military adventures were cut short by an enemy which was no respecter of frontiers or persons; this was the dreaded 'Black Death' (bubonic plague). Whole villages and towns were wiped out and the country did not make up the population loss for two centuries. England, France, and Italy took the full brunt in 1348 and for five years were trying to recover. But by 1355 the hardships of the past were a memory, and Edward once again invaded France. However, he had hardly got the campaign under way before news from Scotland brought him home again. Meanwhile the Black Prince remained in France and plundered the country from Bordeaux through Langeudoc.

This was the policy of 'havoc' and it was continued throughout the following year. The theory of havoc was to avoid conflict but to march burning, slaughtering, and devastating through as wide an area as possible, hoping thereby that general misery would cause the opposing government to submit. As the opposing government was indifferent to the sufferings of peasants there was little constructive about the policy. Edward III was fully in agreement with this futile and callous policy and used it himself when he threw back the Scottish invaders.

But the Black Prince eventually found himself in an awkward position. His army of 7000 was confronted by 40,000 Frenchmen near Poitiers. If the French had possessed any battle acumen they would have realized that the best policy was to besiege the English position, on a twenty-foot high hill, and starve them into surrender. Unfortunately for themselves they decided on a frontal attack, dismounted, with the exception of a small cavalry vanguard. The first two attacking waves were repulsed and fell back on the third. The Black Prince spotted the general confusion and charged on to it. The result was total victory (September 1356) in which the French casualties were ten times greater than the English, and the French King and his son were taken prisoner. The hill of Maupertuis, on which the Black Prince had posted himself, had been protected by a tall hedge and a ditch; these obstacles had sufficed to break the French dismounted attacks. In view of this it is not surprising that permanent fortification such as castles were valued so highly.

However unfortunate the French government had felt themselves before, their condition now became desperate. The barons were unrestrained, disbanded mercenaries roamed and plundered at will, and a general revolt of the peasantry broke out, known as the Jacquerie. In 1359 Edward III crossed the channel with a huge army and ravaged a vast area of country. The French had no alternative but to accept a humiliating peace in which they ransomed back their king and had to yield the territories once linked to England by the marriage of Henry II and Eleanor of Aquitaine.

The Scots obtained slightly easier terms, but had to ransom King David, and hand over Berwick and Roxburgh.

The fortunes of England and Edward were at their height, but this state of affairs was not destined to last long. Apart from Brittany, France remained peaceful, so no action was required there; but internal dissension in Spain offered an opportunity Edward could not resist, and he sent a huge army over the Pyrenees to restore Pedro the Cruel, who had been deposed by Henry of Trastamara. Edward's motives for interfering in this expensive and distant theatre are difficult to understand; he had an alliance with Pedro but was not committed by it; and it seems probable that his main reason for this rash intervention was that the French supported the other side. The expedition ended in failure, and the Black Prince returned to France a dying man, having contracted an unknown fever. His new French domain took advantage of his weakness to rise in rebellion but in a last campaign, in 1370, he captured the rebel stronghold of Limoges, where three thousand men, women, and children were massacred; only a few knights were spared for the sake of their ransoms. Limoges is held to be the classic example of the hollowness of the chivalric ideal. Although his death did not take place till some years later he was too ill to have any further influence on events.

Edward was soon in little better shape than his son for senile decay set in early, and he was managed by a scoundrel called Lord Latimer and a greedy mistress called Alice Perrers. The only active force left in the realm was his third son John of Gaunt who was driven by one motive only, and that was personal ambition.

When bit by bit most of France had thrown off the English yoke, John of Gaunt took a vast army to recapture the lost lands. But the French, knowing their strength, retired to all available castles and strongpoints, leaving the invader to do what he could with a barren countryside. It was siege warfare in reverse, for the intending besiegers could not amass sufficient supplies to make any investment effective. The campaign was a complete failure. Not unnaturally indignation mounted at home against military waste and domestic misgovernment, for which John of Gaunt was held responsible. The opposition found a leader in the lord chancellor, William of Wykeham, Bishop of Winchester and one of the few statesmen of the time. Gaunt tried to embarrass him by supporting John Wycliffe, leader of the Lollards, who criticized the whole conduct of church affairs. However, as it has been said, 'he who sows the wind may reap the whirlwind' and in this case it was so. From reforming the church it was but a short step to reforming the government as a whole, and soon Gaunt found criticism directed against himself and his father. Edward's corrupt supporters were condemned and fined. Gaunt had the condemnations rescinded, only to find that his own supporters came under attack. At this critical stage Edward III died. His son, the Black Prince, was already dead, and the heir to the throne was a 10-year-old boy.

❋ 11 ❋

Lancaster replaces Plantagenet

Richard II (1377–1399): Henry IV (1399–1413)

ONCE again, as on the accession of Henry III, the thought of a young, helpless boy as King rallied support to the throne. William of Wykeham was left to manage home affairs, and John of Gaunt entrusted with policy overseas. Neither had an easy task. The French had by now turned from mere defiance to taking the initiative and had even raided and burnt the important towns of Portsmouth, Gravesend and Winchelsea. The effective prosecution of the war demanded more money, and hence more taxes. This last imposition might have been accepted before the Black Death and the rise of Lollardy but now its effect was to cause a general rising. 1381 saw a partial revolution led by Wat Tyler of Kent and Jack Straw of Essex. At this, Richard II, now aged 14, displayed a personal courage and resolution never to be repeated. He met the rebel army, impressed them with his sincerity and ability, and granted most of the concessions they asked for.

But the greater lords were not so impressed, and once the rebels had dispersed they had Richard's concessions cancelled; the deceived peasants were rounded up and hanged.

The lesson was not lost on Richard. His actions had been made ineffective because he had acted alone. When he was eighteen Gaunt departed to Spain on another adventurous errand, and Richard took advantage of his absence to promote his own supporters. Unfortunately for him he had reckoned without one of his uncles, the Duke of Gloucester. This ambitious and reckless man collected a party of nobles—including young Henry of Bolingbroke, son of John of Gaunt—who styled themselves the 'Lords Apellant' because they claimed to be appealing against the treason of the King's ministers and favourites. The swiftness of Gloucester's treachery gave him an initial success at Radcot Bridge, where Robert de Vere, Earl of Oxford, tried to oppose them with a few locally raised levies. The occasion is notable for the fact that de Vere escaped by swimming the river in full armour. This feat did him little good for he died soon after in exile. There were other occasions in history when men tried to escape by swimming in armour but no one but de Vere seems to have reached the other side.

Gloucester then summoned a Parliament which outlawed Richard's principal friends and ministers. But within a year the self-seeking ambition of the Lords Apellant had turned popular feeling against them. Richard took advantage of having passed his twenty-first birthday to dismiss Gloucester and his party, replacing them with William of Wykeham and other trusted ministers. Eight years of moderation and prosperity followed but unfortunately Richard then decided that he was now strong enough to revenge himself on the Lords Apellant. He displayed considerable skill in tackling this problem. First he won the trust and affection of two leading members, Mowbray and Bolingbroke. With these two on his side he had no problem over banishing Warwick, executing Arundel, and murdering Gloucester. His revenge on Mowbray and Bolingbroke took a more subtle form. Having made both men Dukes, he stirred up trouble between them, and induced them to fight in a tournament to settle the matter. As they were preparing to begin, Richard intervened and announced that he had banished Mowbray for life and Bolingbroke for ten years.

This dramatic success went to Richard's head, and he was soon behaving like a tyrant. Money was raised by forced loans, dissension was stifled by the presence of a small standing army, and the constitution was generally disregarded. His most unwise step was to seize John of Gaunt's vast estates: these were the due of Bolingbroke in spite of his banishment, and the arbitrary seizure aroused sympathy which Bolingbroke was later able to exploit.

The reckoning came when Richard took himself and an army to Ireland to restore English authority. The pent-up discontents of the last few years were now freely voiced, and at this critical moment Bolingbroke landed, ostensibly to claim his lost estates but in reality with larger ambitions. He was rapidly joined by other powerful northern nobles, and, with the opposition disarmed, was soon exacting revenge by hanging some of the King's ministers.

On hearing the news Richard made haste to return but fate treated him most unkindly. For four weeks he was storm-bound in Dublin, and during that time his loyal supporters in England lost heart. He reached Flint castle but was immediately besieged. Resistance was useless and he surrendered, trusting that Bolingbroke had no personal feelings against him; they were cousins. But Bolingbroke was determined on the crown and had no scruples about the steps he took to obtain it. Richard was forced to abdicate and the prior claims of Edmund of March were set on one side; Bolingbroke became King.

Unlike those of Edward II's murder, the details of Richard II's end are not known. As he was sent to Pontefract Castle, it seems likely that the murder took place there. It is possible that Bolingbroke would have been satisfied with keeping him in close confinement had events allowed this, but the insurrection that broke out shortly after his usurpation alarmed him, and in the general slaughter that followed the crushing of the rebellion Richard was probably included. At all events, he was never seen alive again.

Siege warfare took some notable strides forward during Richard's unhappy reign. Guns were already in use although they were both unsafe and inaccurate; Richard owned nearly 100. The advent of small arms, and even deadlier weapons, was signalled by the use in 1381 of rockets with nails in their heads. Very gradually the importance of the mine was receding against the development of the missile.

An incident of 1388 recounted by Froissart gives yet another illustration of the weakness of doorkeepers—a frailty that may also be noticed in much later periods. The subject of the story is Artigat Castle. Two 'varlets' were briefed to get themselves jobs in the town and by that means become well known. On the day appointed for the capture six other varlets went to the back of the town and scaled the outer walls. Then they joined the two better-known varlets outside the inner gatehouse. At this point 'the two varlets called out to the porter "Sire, open the door, I have brought you of the best wine that ever you drank, which my master hath sent you, to the intent that you should keep your watch the better." And they who knew right well the varlet believed . . . and opened the door; and then he whistled and the six stepped forth and entered in at the door, and then they slew the porter so privily that none knew thereof.'

In this way the town was captured, but the castle remained untaken. But among the 'men of the town sitting drinking or else in their beds' was the castellan. He was captured and displayed before the garrison of the castle. The ultimatum was then delivered: either the castle was surrendered or the castellan would be executed in full view. At the threat of seeing her husband decapitated his wife yielded up the keys.

The unusual feature of this ancient incident is the fact that the chief actors in the plot were 'varlets'. When such incidents occurred the performers were usually knights dressed to look like peasants. Originally varlets were apprentice esquires who were usually non-combatants, but later the term became debased to mean 'rascal' or 'menial'.

* * *

The key to understanding the tangled history of the Wars of the Roses, which tore England in pieces between 1454 and 1485, is to be found in the deposition of Richard II. By this act Henry Bolingbroke not merely usurped the crown, and presumably arranged the murder of his king, but he also overrode the lawful claim to it of Richard's heir presumptive, Edmund Earl of March. Edmund was descended from Lionel of Clarence, Edward III's second son and his daughter, Anne, married into the family of Edmund of York, Edward's fourth son: and from this alliance sprang the Yorkist challenge to the descendants of John of Gaunt, Edward III's third son. This dynastic struggle took some time to come to a head. Both Henry IV and Henry V contrived to avoid a major breach although the threat to their position was latent throughout their reigns. It was only under the incapable Henry VI that the pent-up rivalry of Lancaster and York burst into open civil war that continued intermittently for thirty-one years and is known as the Wars of the Roses.

The reign of Henry IV lasted for fourteen troubled years. He had his first narrow escape two months after his accession, when a plot to seize him at Windsor Castle was betrayed.

He was celebrating Christmas and had, somewhat unwisely, omitted to keep a substantial military force with him. The news of this reached the Earls of Kent, Huntingdon, and Salisbury and they collected a force of 400 lances at nearby Kingston; on January 5th, 1400, they launched their attack.

Unfortunately for the insurgents Henry had fled, warned by a traitor a few hours before. The rebels (or loyalists perhaps) entered through a postern but finding the birds had flown and the game was up, escaped to the west. But Henry was not a man to take the same risk twice. He sent troops in pursuit, and captured the conspirators in Cirencester. Their execution was summary, and without trial.

But within months Henry had fresh trouble on his hands. Owen Glendower, a Welsh prince and soldier of genius, took up Richard's cause, believing, like many others, that the latter was still alive. He made himself master of North Wales and raided into England as far as Shrewsbury and Worcester. Attempts to counter-attack either met disaster such as at Pilleth, near Presteigne, in 1402, or lost impetus when Glendower retired to the inner fastnesses of Snowdon.

The battle of Pilleth took place on a very steep slope. The Welsh poured down on the English, who seem to have been caught unawares at an enormous disadvantage. Presumably the Welsh had silenced the English scouts, if there were any.

The Scots were not slow to join in. An army commanded by the Earl of Douglas invaded Northumberland, while in the south French ships raided the English channel ports. Nothing could be done against the French but the Scots were brought to battle at Homildon Hill in 1402. This was a victory for the Percies of Northumberland rather than for Henry and the fact was shown by the valuable Scottish prisoners they took. The proud, warlike, and grasping Percies were overjoyed at the thought of the enormous ransoms they would extract from such prizes as Murdoch, Moray, Orkney, and Douglas, but this jubilation was changed to fury when Henry, desperate for money, and not daring to ask Parliament for it, claimed the prisoners and their ransoms for himself.

A crisis developed quickly. The Percies decided to displace Henry and opened negotiations with Glendower. Douglas was released on the understanding that he would support the revolt, and the French also offered their assistance.

The decisive battle took place at Shrewsbury in July 1403. Henry was swift and purposeful. Hotspur, son of the Earl of Northumberland, and one of the most redoubtable of the Percies, was killed by an arrow (not, as Shakespeare puts it, in personal combat with the future Henry V). By rapid marching Henry IV had won the battle before Glendower had arrived.

The Earl of Northumberland had not been present at Shrewsbury, but not through lack of warlike spirit. He was fined heavily for his part in this rebellion, but two years later stirred up a fresh one. The new venture was frustrated by unfulfilled promises and Northumberland had to escape to Scotland. Two years later he decided the time was opportune for a further move. On this latter occasion he misjudged the situation and was killed, fighting gallantly at the age of 70, in the desperate battle of Bramham Moor.

Glendower's activities made Wales an uncomfortable place for the English at this time. Harlech held out for many months but was obliged to surrender when its garrison had been reduced to 21. Glendower then garrisoned it afresh with his own men and left it in charge of Edmund Mortimer. Unluckily for the, latter he too was besieged when Glendower's armies were driven back to the hills. The strength of Harlech was demonstrated in 1408 when neither mine nor engine could break through, but the weakness of isolated fortifications was also clear when Mortimer, like most of the garrison, died of exhaustion and lack of food in the following January. Harlech surrendered but within the century it would endure the longest siege in English history.

In 1409 Henry's health began the slow decline that eventually led to his death from leprosy in 1413. Fortunately for England the country possessed a number of able Parliamentarians during these four years, and with the exception of Wales, where Glendower still held the inner lines, there was no trouble from bordering countries.

Scotland was temporarily without a king. In 1406 the heir to the throne was captured by an English ship while voyaging to France and was kept as a hostage at Windsor—the Scots King dying of grief at the news. France was bitterly divided by civil war, in which both factions sought English help, and was quite incapable of causing external trouble. Although the French could not know it, they were sapping their own powers of resistance against an onslaught that would soon be launched by Henry IV's talented son.

* 12 *

War in France

Henry V (1413–1422)

HENRY V was 25 when he ascended the throne, but in the previous ten years he had gained valuable experience of war, men, and affairs. He was popular, sensible, disciplined in habits, and purposeful. The latter quality drove him to engage in exhausting though successful wars in France, and to pursue a policy of strict religious orthodoxy at home. Without his foreign successes his determination to stamp out Lollardry and any similar departure from orthodox religion might well have stirred up deep resentment. As it was, the energies that could have produced domestic turbulence were directed overseas. Here, indeed, his ambition went far beyond that of his ancestors. His intention was not only to reconquer all the lost French provinces but also to wear the French crown. And the lost provinces, in his view, included not only Aquitaine but Normandy. The French tried to buy him off with a bride, a huge dowry, and Aquitaine, but their offer was rejected.

His determination to produce an overwhelming force was such that he even pawned the crown jewels. His force amounted to 10,000 men, of whom three-quarters were archers. The remainder were lances (who required from two to four horses), smiths, painters, armourers, tent-makers, fletchers (arrow-makers), and bowyers. In this army the heralds, although few in number, were highly esteemed, for they had important duties at sieges; they had the knowledge which enabled them to stipulate who was entitled to strike the first blow, who was the chief captain of the opposition, and what the correct procedure of surrender entailed. A beaten and starving garrison was not allowed to say 'We have lost' and submit to punishment; the whole ceremony must be conducted with due protocol or the fight would continue. It would have been fatal for a surrender party to emerge without the right armorial bearings, i.e. grabbing the first shield that came to hand and raising a white flag above it. Henry would have set his heralds to work, and the imposture would have been discovered immediately.

The army's first engagement was at Harfleur, an event which is famous without being well understood. The French had every opportunity to oppose Henry's landing which was over marshy, stony, and rough ground, but they did not. The siege began on August 17th, 1415.

The town was two and a half miles in circumference and surrounded by deep ditches filled by water from the Seine. In addition the Lézarde flowed through the town. The garrison was small but efficient; jutting emplacements gave them a good field of fire.

Henry surrounded the town and proceeded to batter the walls with both guns and engines; he attempted two mines, but both were frustrated by countermines. Success came into view when he succeeded in breaking the Lézarde bridge and flooding the town, but Harfleur's troubles were nothing to his own. The French made a series of irritating sorties and far worse damage was done by the dysentery which broke out among the besiegers. In spite of, or perhaps because of, strict discipline desertions were frequent. Harlots were warned not to approach within three miles of the English lines and orders were given that, if the warning was disregarded, their left arms would be broken before they were ejected from the camp.

By September 18th, Harfleur, with disease, a contaminated water-supply, and considerable battle casualties, offered to surrender if not relieved within four days. Henry was pleased to accept—his own force was down by one third—and on September 22nd he received the keys of the town.

Having garrisoned Harfleur he was in some doubt about the best move for his depleted army. He chose what appeared to be the rashest course, for he resolved to march from Harfleur to Calais across country swarming with newly-organized French forces. In the opening stages it was possible to outmanoeuvre them by speed (for he had mounted the entire army on commandeered horses), but when he crossed the Somme after some delays he found a French army of some 30,000 drawn up and waiting for him near Agincourt.

However, the French, with extraordinary stupidity, massed themselves together in three dismounted columns and advanced over newly-ploughed sodden fields. Henry met them on a narrow front and deployed the archers on the flanks. By their approach in heavy armour, over mud, the French placed themselves at maximum disadvantage, and were slaughtered freely, first by the archers, and then by the lightly-clad English infantrymen, most of whom were bowmen who had now run out of ammunition. It was less of a battle than a form of French suicide, for the English lost only 200 in accounting for 10,000. It was also the writing on the wall for armour, for the better armoured were the chief sufferers and formed the major portion of the French casualties.

Apart from a few ransoms, and a lot of glory, Agincourt was a barren victory. Henry moved on to Calais, and thence to England at the end of the year.

Henry did not open the second campaign until the summer of 1417. On this occasion his army was nearly half as large again, numbering some 16,000. Preparations had been thorough. As an example we note that in the previous February sheriffs had been ordered to have six of the wing feathers plucked from every goose except breeders, and sent to London for the fletchers.

The memory of Agincourt stopped the French opposing him in the open field, but the Norman towns put up a worthy resistance. Among them were Caen and Cherbourg. Caen had walls seven feet thick, twelve gates, and thirty-two towers. It also had an excellent system of defensive ditches. Against this were two weak points in the shape of abbeys on the east and west; both overlooked the town. The French would have burnt them had not the Duke of Clarence moved in first. He was able to do so because a monk had preferred to betray his countrymen rather than see the abbey destroyed; as this was before the rise of French nationalism, and the choice was between two warring factions, it should not be compared with the betrayals of Ely.

After the first assaults on the town had failed, mining and

cannon fire were intensified. Incendiaries were extensively used. Mine-detecting was accomplished by the Middle-Eastern device of leaving bowls of water on the ground; any underground activity would be revealed by quiverings on the surface of the water.

The Duke of Clarence was the first upon the walls; the Earl of Warwick accomplished a similar feat by being the first to scale a great tower called Little Castle. Other enterprising warriors were less fortunate. Sir Edmund Springhouse slipped off a scaling ladder, fell into the ditch, and was roasted alive by burning straw which the French threw on him. Henry's knightly reputation is somewhat diminished by the ruthless behaviour of the English after the fall of the city; it is said that by permitting excesses on this occasion he hoped to destroy the French will to resist in other towns.

Cherbourg was an extremely difficult siege. As the bridge across the harbour was broken, Gloucester had to camp on shifting sands, which made mining impossible. His camp was under constant fire from the walls, and attempts to build earthworks failed through the intensity of this attack. Some palisades were erected with great difficulty during darkness by soldiers who had to swim pushing them forward; but they were soon broken or burnt when dawn came. The Earl of March, showing great determination and daring, pushed a 'sow' right up to the walls but this too was soon destroyed. But eventually starvation and attrition won the day for the besiegers after a five-month struggle.

Henry appointed himself Duke of Normandy, and gave the state a more orderly government than it had experienced for many years. But Henry knew that it was one thing to call himself the Duke and another to control the duchy; the key to control lay in the possession of Rouen.

Rouen was well-prepared. Ditches had been deepened, caltraps sown by the thousand, pits dug, and earth banked high inside the walls. There were five gates on the landward side, and sixty towers in between. Each tower had three guns, and between each tower were eight smaller guns and catapults. The garrison numbered about 20,000. Our knowledge of the siege is fairly complete for we have three contemporary accounts, one by an eye-witness, John Page.

Henry built a fort in front of each gate, and connected them with palisaded trenches. Then he closed the river to navigation by chains and booms. His next move was to gain control of the strategic Pont de l'Arche, six miles up river, as Rouen, which was then on the northward side of the river only, could be approached from behind.

The attack on the Pont de l'Arche was preceded by a demand for its surrender by Sir John Cornwall to the Sire de Granville. Granville refused and Cornwall then wagered his steel helm (which was worth 500 nobles) against Granville's best charger that he would cross the river.

The first attempt was a night crossing on pontoons made of hide stretched over wicker work; these had been brought over from England. It was accompanied by a diversionary feint attack three miles downstream, and in consequence 5000 crossed the pontoons without a single casualty.

But as the siege went on, and Rouen held out, relationships became decidedly less chivalrous. The duels and challenges, jousts and tournaments fell into abeyance, and besieged and besieger settled down to a sullen fight to the finish. The English began hanging their prisoners where the townspeople could see them; the French retaliated by drowning theirs in sacks in the Seine. When 12,000 non-combatants were expelled Henry refused to allow them through the lines; their condition was soon desperate.

Henry was as ruthless with himself as he was with others. He exposed himself to rain, sleet, snow and, of course, enemy fire. At one point he dressed up some troops to look like a relieving column of Burgundians, hoping to provoke a sally from the garrison, but the stratagem did not work. By early December, with the siege now seven months old, Rouen was desperate. Five thousand were said to be dead from starvation and there was an ugly rumour of cannibalism among some of the remainder. A desperate sortie of 2000 from each gate accomplished nothing. Christmas came, and in the one-day truce grudgingly offered by Henry, and even more reluctantly accepted, the dying between the lines were given a small quantity of food. As they had lain in continuous rain on flooded ground for some two months they were not greatly affected by this demonstration of Christmas goodwill.

No small part of the English success was due to the Welsh and Irish, who accompanied the English cavalry on their raids into the countryside. As noted before, the Welsh usually had courage, daggers, and bows if they were lucky. The Irish kerns rode bareback with one foot and leg naked; their weapons were javelins and knives. Unsupported, such troops were usually easy victims, but as units of a disciplined force made magnificent skirmishers.

Eventually it was decided that if relief did not arrive the city would surrender on January 13th. Needless to say, there was no relief and Rouen capitulated. Subsequently they may have regretted their submission for many died of the effects of privation, and those who survived faced the task of paying an impossibly large ransom.

All through his invasion Henry had been assisted by the bitter rivalry raging between Burgundians and Armagnacs, which prevented any concerted effort being made against him. But in 1419 the balance of power tilted further in Henry's favour. In that year Duke John of Burgundy was invited to a peace conference at Montereau where he was treacherously murdered by the Armagnacs—an act of folly that quickly brought about an Anglo-Burgundian alliance. As the Armagnacs backed the French King and the Dauphin, active Burgundian support was a valuable aid to Henry's designs on the French crown. With this welcome, but not necessarily permanent, alliance he completed his conquest of the greater part of northern France, including Paris, and by 1420 his fortunes reached their peak. His success was marked by his marriage to Catherine, the daughter of the King of France, Charles VI. In fact, Charles was imbecile and his Queen whole-heartedly detested their son, the Dauphin. It was not difficult, therefore, for Henry, backed by his victorious army, to have himself formally accepted as the heir to the French throne, which he sincerely believed he was.

But, as part of the peace treaty, Henry had to subdue the Armagnacs. At Montereau he occupied the town within two weeks but was faced with the prospect of besieging the castle for much longer. He therefore adopted a simple strategy. When the town had fought, the wives and children of the leading defenders had been sent to the castle for safety. Henry promptly sent a message to say that his captives would all be hanged in front of the castle unless the garrison surrendered. He thereupon built a gallows, appointed a time, and waited for results. The doomed men asked for a last sight of their wives and children, and firing stopped while these leaned over the battlements and waved good-bye. Then they were hanged.

Eight days later the garrison had to capitulate and to its surprise was treated with mercy and courtesy.

The occasion is also memorable for the hanging of one of Henry's favourite grooms. The man had accidentally killed a knight, or if the death had not been a pure accident it was certainly exculpable. But not for Henry. Deeply though it grieved him, he had the man hanged on the gallows on which swung the bodies of his former enemies. The crime was not, of course, murder, but killing a man of a higher rank.

Melun proved an altogether tougher assignment, the siege lasting from July 13th to November 17th. On this occasion Henry deserted his usual practice of waiting for starvation to take its toll and instead captured an outpost in a bloody assault. After this initial venture he waited, although the Duke of Burgundy went forward and was repulsed with crippling losses. The fight then became a normal siege of attrition apart from one feature. Mine and countermine met so often that ultimately there was a second battlefield underground. In this subterranean arena large forces would meet by appointment, the combat being announced by trumpets, as if for a tournament. Even Henry himself fought underground (Plate 3).

When Melun fell, its commander, Arnold Guillaume of Barbazan, was imprisoned at Château Gaillard, but was rescued ten years later.

Peace was short-lived. Although the Scottish King was a prisoner in England this did not prevent the Regent, the Earl of Buchan, from aiding the Armagnacs. In 1421 the latter gained a sweeping victory at Beaugé, where the Duke of Clarence, Henry's brother, was killed.

Henry responded almost immediately, but the elimination of the Armagnac strongholds proved unexpectedly difficult. Dreux capitulated on August 8th, 1421, partly because of starvation, but Meaux, which was invested on October 6th, 1421, held out for seven months. It was the most difficult of all Henry's sieges, and the one in which he contracted the illness which brought about his death, in his thirty-fifth year.

Meaux was well-known on account of the activities of its commander, De Vaurus, otherwise known as the Bastard of Vaurus. He massacred every Burgundian he could lay his hands on, and for good measure classified the English as Burgundian. When short of either he attacked anyone within reach. His victims were hanged on an elm tree.

The town lay on the end of a peninsula which jutted out into the Marne. The peninsula had been cut to make an island of the tip but the south bank had been cut again further along making another island of the district known as the Market. Henry positioned the Duke of Exeter on the north, and March and Warwick on the east and south respectively.

The garrison numbered about 3000, and seldom can three thousand braver and more desperate men ever have been besieged. Mine was met by countermine, attack by counter-attack. As soon as a wall was breached the defence would concentrate its force, drive back the attackers, and repair the gap Henry had to contend with floods on the Marne and also deserters; one of the latter was the valiant Sir John Cornwall who had seen his young son killed in front of him. But discipline was never relaxed. A soldier caught stealing a pyx was promptly hanged, and Henry was everywhere to see that his strict orders were obeyed to the letter.

After five months the Dauphin made an unsuccessful attempt at relief. Only one column got through, and this consisted of 40 men led by Guy de Nesle. It passed through the English lines by killing the sentries, and scaled the walls on ladders which the garrison had covered with bed sheets so that they would not show up against the white walls. At the very moment of success de Nesle fell and was captured as he lay bruised and in full armour at the bottom of the ditch.

The failure of the relief threw the townspeople into despair. Their morale was already low because of an English mine that they had been unable to locate and counter and, now that the hope of outside help had gone, they planned to burn the town and move into the Market. In the event the town was not burnt, for the plan had been divulged to Henry by a citizen who did not wish to see his house go up in flames, but the garrison escaped into the inner stronghold.

Two events then intensified the misery of the defence: first the English built a tower with a drawbridge and used it for assaulting the gate, secondly the besiegers captured the town's mill and made the grinding of corn impossible. But the greater the misery the stronger their morale. A call for surrender was met by a brief and indelicate answer, and a successful sortie killed a party taken by surprise in a nearby meadow, But Warwick had now got a sow to the walls and was working in. Sir Walter Hungerford did the same in the west but was disconcerted when the defence opened a breach from within and defeated him in hand-to-hand fighting. In the last desperate battle, with the walls in rubble and both sides fighting with anything that came to hand for a weapon, the conflict was so fierce that the siege might have been just beginning. Eventually it was the sight of an approaching assault tower mounted on two boats that convinced the inhabitants that it was hopeless to resist further.

Henry's peace terms were surprisingly lenient. The Bastard of Vaurus was beheaded, and his body displayed on his own favourite elm-tree; the English and Welsh who had fought inside were also executed, but the remainder were merely made prisoners of war. It was May 2nd, 1422. In that moment of heady triumph, at the height of his glory, Henry was already a dying man. He died on August 31st. Had he lived six weeks longer he would have inherited the French crown, his life's ambition: but whether the union of the two kingdoms, if he had lived to have established it firmly, would have lasted, or even proved viable during his lifetime, or benefited anyone, is a debatable matter.

✲ 13 ✲

The Wars of the Roses

Henry VI (1422–61 & 1470–71)
Edward IV (1461–70 & 1471–83)
Edward V (1483) : Richard III (1483–85)

TO have a young boy as King was not a new experience for England but this child was not yet twelve months old. On his death-bed Henry had appointed Duke Philip of Burgundy as ruler of France, and nominated his younger brother, Humphrey of Gloucester, as Regent of England. This latter was a surprising move for Gloucester was already known to be headstrong and unreliable, whereas John of Bedford was passed over although his qualifications were obvious to all but the dying Henry. However, Bedford had an exacting task in France, where he proclaimed Henry VI King when the old French King Charles VI died. The fact that the French promptly declared for the Dauphin, an idle and unattractive young man, meant that France had two kings, neither of whom had any real qualifications for the post. Although the English had gained most of Henry's objectives there was no real peace, and the war had now gone on so long, with bitterness and retaliation on both sides, that it seemed it might last for ever.

However, an act of incredible stupidity by Gloucester, who bigamously married a Burgundian heiress and tried to occupy her French dominions, damaged the English alliance with Burgundy, in spite of Bedford's protests that he was completely at odds with Gloucester. The crisis was staved off by the ludicrous failure of Gloucester at Hainault, after which he deserted his bride in bigamy, and returned to England to marry one Eleanor Cobham, whose moral reputation matched his own. But the effects were seen when Bedford sent an army of some five thousand, led by the Earl of Salisbury, to take

Orleans. This siege began on September 12th, 1428, by which time the French had already destroyed the surrounding vineyards and houses, as well as set up two strong forts. The English force was too small to mount a proper siege and had to be content with blocking the main entrances. This policy was anything but successful for Orleans was large and had numerous means of ingress.

Salisbury opened his attack but soon had half his face removed by a stone; he survived the ghastly wound eight days. Suffolk succeeded to the command but had neither the reputation nor skill of his predecessor. By the spring the besiegers were as short, if not shorter, of food than the besieged. In consequence the Duke of Bedford sent a large convoy under the command of Sir John Falstaff (not to be confused with Shakespeare's Sir John) which was intercepted by a detachment of French and Scots from Orleans, numbering, it is said, 8000. The English, reported to number no more than 1500, pulled their wagons into a circle and fought behind its protection. The ensuing battle of February 12th, 1429, has become known as the Battle of the Herrings, for the consignment included the Lenten fish supplies. The victory went to the English and illustrates the simple principle that attacking a defended position without weapon superiority needs a more than 6–1 ratio. An enclosure made of a few baggage wagons may not seem much of an obstacle but this was what constituted the laagers from which the Afrikaaners were able to beat back hordes of Zulu warriors in the nineteenth century.

The blockade of Orleans then became tighter, although it was a long way short of being a stranglehold. At this critical stage France found an unexpected leader and mascot in Joan of Arc. The simple peasant girl who saw visions and heard voices aroused fresh flames of resistance in the French, who saw her as a semi-divine saviour. Outside Orleans the besiegers were in little better plight than the besieged, and the fact that the Duke of Burgundy had quarrelled with Bedford and departed with his troops made the task of the English army seem hopeless as well as arduous. However, it was not sufficiently arduous to make them obey a condescending note from Joan of Arc which told them, with critical comments, to surrender the keys of all captured cities and go home. Joan thereupon set out with a relief column. She was with difficulty persuaded from approaching through the middle of the English lines, and instead set off in boats up river. It was stormy and they made little headway. She was urged to discontinue fighting against such a wind. 'It will change,' she said, and change it did. Supplies were landed six miles below the city, the garrison made a diversionary sortie, and Joan rode in through the east gate.

From the moment she arrived in Orleans French morale soared. Joan initiated one attack after another, all of which were successful. She was not unscathed, however, being wounded twice, though not seriously. Before the final rout of the besiegers William Glasdale, a tower commander, scornfully told Joan to get back to her cows. To this she replied, 'Your men will retreat but you will not go with them'. A week later Glasdale was crossing a wooden bridge when it was struck by a missile and collapsed into the stream. Glasdale was drowned. This incident, darkly ominous to the mediaeval mind, was the last straw for the English who left their forts and retreated. Some managed to surrender, others were killed or drowned. In eleven days Joan had clinched victory over an army that had ravaged France for eleven years.

The tide had turned in earnest. Fresh levies joined the Maid of Orleans, and before long the English forces suffered another heavy defeat at Patay where Talbot, the English commander, was captured.

At this point Joan would have returned to her village, but was not allowed to do so. Somewhat unwillingly she went on, but failed to take Paris, which was defended for the English by the Burgundians. However Soissons, Laon, Beauvais, Senlis, Compiègne, and Troyes all fell to her army and the English holding in France was now limited to a defensive triangle. Compiègne was soon under counter-attack and in trying to relieve this siege Joan was wounded and captured. Her Burgundian captors sold her to the English, who regarded her as a witch and a heretic. She was tried by a French court, condemned, and burnt alive in the market place of Rouen.

The deed was not only cruel and shameless; it was also unwise. Joan was no longer a defeated captain who had temporarily inspired morale; she was a martyr who would now inspire and guide the French until the last English invader was thrown out of the country. During this time the Burgundian alliance was beginning to wear thin, until it was finally dissolved in 1436. As John of Bedford, the mainstay of the English struggle for France, had died in the previous year matters began to look black for the invader. Paris surrendered to the French King, and the English were soon reduced to defending a narrow strip in Normandy and Maine.

But there were still twelve years of French war to follow. The hero of this phase was Talbot, Earl of Shrewsbury, who hung on with grim obstinacy. A strong peace treaty at home managed to achieve a three year truce but in 1449 fighting broke out with greater fury than ever. The cause was local. The Norman garrisons had mutinied from lack of pay and sacked Fougères in neighbouring Britanny. Charles VII, the French King, declared war and to everyone's surprise swept the English before him. Rouen fell, betrayed from within, and the last English footholds left in Normandy were Cherbourg, Harfleur, and Caen. Their survival was short-lived. An English defeat at Formigny (April 1450) brought about the surrender of Caen and with its fall Normandy passed finally into French hands. Attention was diverted from this disaster by events at home: the Earl of Suffolk, one of Henry's trusted ministers, was murdered, and this was followed by Jack Cade's rebellion. Cade did not last long, but Henry refused to reform the conditions which had caused the uprising. Instead he attempted to direct attention overseas by what he and his advisers hoped would be a successful French war. Talbot, Earl of Shrewsbury, landed at New Bordeaux with a mere 500 men, took the city and had some further successes. Unfortunately for the expedition he was killed in the following year (1453) at Castillon. Bordeaux was starved into submission, and Calais was all that remained of the mighty French Empire.

But the end of foreign adventure did not mean that the English appetite for warfare was glutted nor that the leading families had acquired peaceful tastes. On the contrary the lack of opportunity overseas meant that the barons were forced to concentrate on their family feuds and interests, to serve which there were available numbers of unemployed and experienced soldiers returned from the French wars.

The origin of the Wars of the Roses, as we have seen earlier, lay in the murder of Richard II by his cousin who became King Henry IV. Like his father, Henry V was an able and successful King suppressing without much difficulty any challenge to his authority; but when Henry VI eventually ascended the throne after a long minority, and it was clear that he was far from able or successful, the existence of a legitimate grievance gave a fine opportunity for ambitions that were based on local aversion rather than national and legal considerations. Family hatreds subsequently kept the conflict going when continuance was disastrous to the nation.

By 1453 it was patent, even to the most tolerant, that Henry VI was insane. Had he stayed so there might have been hope for the future, for his cousin, the Duke of York, administered the realm competently and loyally; but after eighteen months Henry recovered, and elevated once more his favourite, the unpopular Somerset. The opening battle took place at St Albans in May 1455, when Somerset was killed and Henry taken prisoner. York became uncrowned King. Henry's wife Margaret feared, however, that York would hold on to his new-won power and deprive her infant son of his heritage. Accordingly she enlisted all the support she could find. After a few skirmishes her forces gained a devastating victory at Ludford. York escaped to Ireland, while Warwick took refuge in Calais.

But Queen Margaret exploited her triumph too harshly and too soon. Warwick took advantage of the general disaffection to return and seize London. From there he was able to advance and beat the royal forces at Northampton in 1460; once more Henry was taken prisoner, although Margaret escaped. The occasion was notable in that the casualties were mainly among the leaders of the Lancastrians, for the underlings were spared.

The Duke of York was so embittered that he was no longer prepared to do duty for the King; and he put forward his own claim to the throne which was, in fact, stronger than Henry's.

Meanwhile Margaret had been gathering forces in the north. York marched north to meet her but had underrated the opposition. Outside Sandal Castle, near Wakefield, he suffered overwhelming disaster and was beheaded. However, his 18-year-old son was in the west to continue the struggle. Margaret's army pressed south, and secured another crushing victory in 1461 at the second battle of St Albans. Had Margaret marched on to sack London the course of history might have been different. However she was restrained by her husband and the delay proved fatal. Young Edward of York, having just beaten the Lancastrian's Welsh supporters at Mortimer's Cross, raced to London and reached it as it was on the point of capitulation.

Within a few days the entire situation had changed. Margaret's army had dispersed on plundering forays: Edward of York was therefore able to advance without much difficulty. He pursued the royal forces to Yorkshire and caught them at Towton, near Tadcaster. The ensuing battle was ferocious. It lasted from dawn till dusk of March 29th, 1461, and was fought on a hillside in a blinding snowstorm. Many of the Lancastrians who escaped death on the battlefield were drowned in the River Cock which ran behind the battlefield. Henry VI and Margaret escaped and fled to Scotland.

Although Edward was a highly skilled soldier he was idle and incompetent in time of peace. Once he had assumed the crown he handed over most of his duties to the Earl of Warwick, the 'Kingmaker'. The latter spent the next three years trying to crush the last remnants of opposition, which were centred in certain Welsh and Northumbrian castles. The year 1462 saw Bamborough, Alnwick, and Dunstanborough captured by Warwick only to be lost to Margaret twice through treachery. After the battles of Hedgeley Moor and Hexham in 1464 the Lancastrians were beaten and in June Bamborough was retaken by Warwick with heavy artillery.

At this point, when the Yorkist cause was at its peak, Warwick quarrelled with Edward over the King's secret marriage. Edward was not disturbed but endeavoured to humiliate Warwick, who had become too powerful. However, he miscalculated the latter's resourcefulness, for Warwick allied himself to the treacherous Duke of Clarence and with his help beat Edward's army at Edgecote Field near Banbury, following up the victory with the capture of Edward himself in Buckinghamshire. Later, with rash clemency, they released Edward in 1469, on promise of co-operation and good behaviour. However, promises counted for little in the Wars of the Roses and Edward had soon raised an army that chased Warwick and Clarence out of the country. Warwick, with no other thought than revenge, offered his services to Queen Margaret: and by the end of 1470 this unlikely combination, exploiting local disaffection, had driven Edward from the country. Henry, a mere shadow of a man, was once more made King of England. His reign was brief. Clarence betrayed Warwick and joined Edward. On April 14th, 1471, Warwick was defeated and killed in the Battle of Barnet. On the same day Margaret landed at Weymouth to join him but on hearing the news of the disaster she marched north-westwards. Edward caught her army at Tewkesbury, killed her young son, and took her prisoner. Henry VI was then murdered, presumably on Edward's instructions: and his death marked the end of the usurping line of Henry IV of Bolingbroke.

Edward IV began his second reign in 1471. His position was secure. The only possible claimant to the throne was a boy of 14, Henry of Richmond, who was living in Brittanny, to which his mother had wisely sent him. Unfortunately, Edward did not need other enemies than himself to bring about the downfall of his house. His sloth and his self-indulgence became notorious and increasingly he left public affairs to his younger brother, Richard of Gloucester, who had fought with him and shared his period of exile. Gloucester succeeded in retaking Berwick, which the Scots had held since Towton in 1461. In 1483 Edward's excesses brought him to a premature grave; his heirs were Edward aged 12, Richard aged 9, and five daughters. These unfortunate children were the pieces in a power game now to be played to a finish. On one side were the Woodvilles and their kin who were related to the late King's wife; on the other side were the supporters of Gloucester. These two factions were already poised for conflict.

The confrontation began immediately. Gloucester captured the two princes on their way to London and assumed the Regency himself. With the boys in his power he proceeded to remove his opponents on the grounds that they were enemies of the young King. This done, Gloucester next 'discovered' that the late King's marriage had been bigamous and, as the corollary of this, that the two princes were illegitimate. It was very simple and very sad: and the best solution to this dynastic embarassment was that Gloucester should take the crown himself—which he did. At the time the young princes were lodged in the Tower. They were never seen alive after August 1483. The details of the murder are unknown and doubtless will remain so, but there was no doubt in the minds of contemporaries as to the man who had given the orders for it.

When Gloucester was crowned as King Richard III, there was only one surviving claimant to the throne, Henry Tudor—still living in Britanny but now a man of 26 years. His claim was slim. It derived from the fact that Henry V's queen, after the King's death, had married Owen Tudor, a prosperous Welsh country gentleman though hardly a match for a royal widow. Their eldest son, Edmund, married Margaret Beaufort, the wealthy Countess of Richmond and Derby, great-granddaughter of John of Gaunt. The offspring of this marriage was Henry Tudor. In more normal times a claim founded on such tenuous grounds could have been safely ignored but it was powerfully reinforced by the widespread discontent bred by Richard's executions and ruthless government. At Bosworth, in August 1485, the crown of England again changed hands on a battlefield that was small even by mediaeval standards. And the Wars of the Roses were at an end.

Among those whom Richard had sent to the block was William, Lord Hastings. He had begun a most interesting castle at Kirby Muxloe in Leicestershire, which he did not live to see completed. This had a wet moat, wide square towers, a strong gatehouse and gunloops. The architect, Couper, had helped design Eton and Winchester, but his talent was by no means limited to scholastic enterprises. The building had not been begun till 1480, and not the least remarkable feature of it is the builder's faith that he could make it defensible against artillery, although sixteen years before the hitherto impregnable Bamburgh had been captured by the Earl of Warwick using only two canons. In the seventeenth century the events of the Civil War were to show that other castles could be defended by gunfire although they were not as well provided with gunloops as Kirby Muxloe. Hastings, or perhaps Couper, had been quick to grasp the fact that new weapons of attack may be converted to defensive uses and can well prove more than adequate for their purposes. (The machine-gun, originally used as an attacking weapon found a different and deadly role in World War I). But, in spite of the imaginative use of gunloops, Kirby Muxloe could never have had more than a nuisance value: superior forces or starvation would have ground it into submission.

But faith in fortresses is not easily destroyed. In 1473 when the Lancastrian cause appeared to be finished, John de Vere, the then Earl of Oxford, attempted a revival. He chose Cornwall for his launching bid, obtained two ships from Louis of France, and landed 200 followers in Mount's Bay. They were disguised as pilgrims but underneath their habits they wore armour; as they were all Lancastrian exiles their piety was practical rather than ethical. On September 30th they claimed entrance to the shrine to make their offerings, but as soon as they were admitted took possession of this powerful water-fortress. From St Michael's Mount they tried to raise a local rebellion but found no support. In time the news of this centre of disaffection came to the ears of Edward IV and he instructed the Sheriff of Cornwall to disperse it. Unfortunately for the Sheriff he was killed in an attempt to attack at close quarters. His successor, John Fortescue, appeared with 4 ships, 9000 archers, and a formidable array of guns, but found it quite impossible to reach the mount, which was surrounded by water for half the day. Gunfire was completely ineffective owing to the height of the target, but starvation reduced the mount six months later in February 1474. The fate of the rebels who had caused so much trouble was imprisonment.

Appropriately, the last siege to be described in this book was also the longest, and perhaps the most influential in its ultimate effects. It took place at Harlech and lasted seven years. Unfortunately, very few details were recorded, and even the incident commemorated by the famous 'March of the Men of Harlech' is now a mystery. Henry Tudor, later to be Henry VII of England, was confined in the castle, and this experience in his formative years may well have contributed to his suspicious nature, scraping frugality, and dislike of baronial power.

In the early stages Harlech was only technically besieged as it was the battle headquarters of a considerable area of North Wales, from which Jasper Tudor, half-brother of Henry VI,

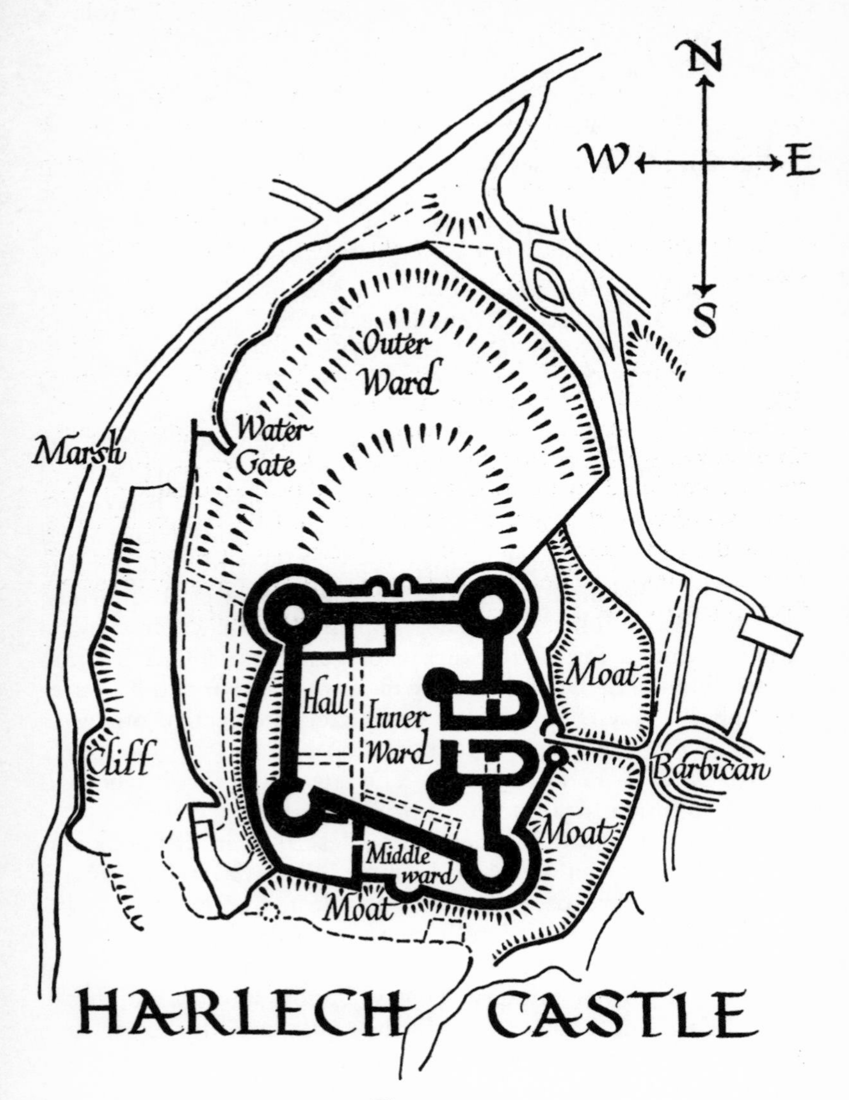
N
W
E
S
Outer Ward
Water Gate
Marsh
Moat
Hall
Inner Ward
Barbican
Cliff
Moat
Middle ward
Moat
HARLECH CASTLE

Figure 21.

harassed the Yorkists with guerilla warfare. When the Yorkists closed in, Jasper escaped to Ireland but he left behind a formidable warrior to be castellan of Harlech. This was Sir David ap Jevan ap Einion who is reputed to have said 'he had once in his youth maintained a castle so long in France that every old woman in Wales had heard of it, and in his old age had held a castle in Wales so long that every old woman in France had heard of it'. It appears that he beat off a few Yorkist attacks with contemptuous ease, and surrendered only when faced with complete starvation. He was granted honourable terms.

Part of the secret of Harlech's success lay in the fact that it was situated on a crag two hundred feet high, in the middle of a marsh. It had access to the sea and was therefore difficult to blockade effectively. It was a concentric castle. A notable feature of its design was the narrowness of the middle ward; an attacker who had breached the outer walls would find himself too confined to be able to mount a proper attack on the inner ward with the right numbers of men. Like a man fighting in the dark, he would never know whether his blows were going to fall on friend or enemy, and would suffer as much frustration from his own side as he would from his opponents.

* 14 *

Conclusion

THE number and vigour of the incidents described in this book may have given the impression that life in mediaeval castles was full of excitement and hazards. There were, as we have seen, occasions when castle-dwellers saw enough action to satisfy the most bloodthirsty, but it would be quite incorrect to think that fighting was a regular event for English castles. Some, such as Guildford, never saw a siege at all; others saw one or two sieges in the course of two or three hundred years.

But the fact that castles did not always have to fight did not mean that they did not have to function. Guildford Castle was built to guard the Guildford Gap, which is a gorge cut by the river Wey between the Hog's Back and the remainder of the North Downs. It therefore controlled one of the southern invasion routes to London, and was of considerable strategic importance although its influence was never exerted directly. During the first hundred and fifty years of its existence it appears to have been an administrative centre and royal hunting lodge; after the year 1250 it appears to have been the main Sussex prison. A notable guest was Sir Adam Gordon, who had led a Robin Hood existence since the Battle of Evesham, but was taken prisoner in single combat by Prince Edward in 1276. Edward obtained a pardon and presented him to the Queen who was at Guildford at the time. Other prisoners were less fortunate, and seem to have suffered considerably from overcrowding until the Sussex prisoners were allowed a gaol of their own at Lewes in Henry VII's reign.

Life in a castle was undoubtedly much like life in a garrison at any period. During settled times the chief enemy would be boredom, and the most important influence on morale would be the weather. The higher ranks would divert themselves with hunting, drinking, gambling, and crude games; the remainder would be occupied with guard duties, quarrelling, routine patrols over a twenty-mile radius, and the establishment of temporary or permanent relationships with any women in the area. Like all soldiers they would alternate between weapon training and domestic chores, would constantly complain to each other that they lived in the worst castle, with the filthiest food, and the most incompetent officers in the country, but would fly at the throat of any stranger who dared utter a word of criticism of it. They would know each other very well indeed —who was brave, who was foolhardy, who was a bit of a liar, and who needed watching when he walked near other people's property. They would despise civilians, although they might envy their more independent existence, and even when they became too old for military duties they would not become civilians but simply 'old soldiers'. Whether miner, archer, infantryman, gynour, or hobiler, they would know, and let others know, which was the truly élite arm. Doubtless there were also élite castles whose defenders felt the pride of the modern fighting man handling the latest and most sophisticated equipment.

Although the main era of siege warfare was over by 1485, this was not the end of sieges or castle-building. The Civil War saw most of the old structures regarrisoned and strengthened, and some of its sieges, such as Denbigh and Corfe, were the equal of anything which had preceded them. In the 1540s when Henry VIII had contrived to make enemies of the Pope, the Emperor, and the King of France, at the same time, he built a series of small castles along the south coast from Deal to Portland. These were intended as an anti-invasion measure but were never tested.

Many years later, when it seemed that Napoleon's army might cross the Channel, a further set of castles was ordered by Pitt. These were the Martello towers, which are well known, if somewhat baffling, to south-coast holidaymakers.

1940, and the threat of a German invasion, saw fortification on a very wide scale, though never equivalent to the Maginot Line or the coastal defences of France. One of its more interesting features was the adaptation of sites and buildings last used many hundreds of years before. Pevensey Castle is the outstanding example. To quote the Ministry of Works Guide: 'With the collapse of France in 1940 and the imminent threat of German invasion Pevensey Castle suddenly resumed its original military purpose of protecting the British coast. In May, 1940, the castle was re-fortified for use as an observation and command post. . . . The towers of the mediaeval castle were fitted up inside for the accommodation of men; "pill-boxes" for machine-gun defence were erected on the keep, on certain of the Roman bastions, and among the fallen fragments of the Roman walls; the main south-west entrance of the Roman fort was closed by a block-house for anti-tank weapons, and an entirely new tower was added to the eastern wall.'

Some of the additions to Pevensey were subsequently removed, but others were left because it was felt that they were an essential part of the castle's military history.

Perhaps the last word on the castle is that it has now gone underground, underseas, and overhead. It is now the deep rocket site, the Polaris submarine, and the military aircraft, one version of which, the B29, was not unsuitably named 'The Flying Fortress'.

The last round in this phase may well be a castle in the air in a practical sense. Space satellites are said to be the ultimate military sophistication, and as such unassailable. So, of course, was Château Gaillard.

Glossary

Adulterine Castles: Castles built in Stephen's reign without royal permission.
Allure: The wall walk within the parapet.
Ashlar: Square stone blocks used for facing.
Bailey: Enclosure around castle; also known as the ward.
Ballista: Early crossbow or sling; the word was used loosely.
Beffroi: Siege tower, sometimes called a belfry. Compare belfry in church towers; the word originally meant 'shelter' and has no connection with bells.
Bombard: Early cannon with range of several hundred yards.
Brattices: Wooden platforms built out from battlements enabling defenders to drop material on attackers below; also known as hoards.
Casemates: Galleries outside the base of the curtain wall. They were provided with arrow loops and were used for defence against miners, battering-ram teams, etc.
Chat castel: Bore or ram.
Crenellations: Battlements. The upper part of walls were divided between *merlons* (stonework) and *embrasures* (spaces). Licence to crenellate meant royal permission to fortify.
Crow: Hook used for pulling down battlements or snatching up attackers.
Curtain: The wall around the bailey.
Donjon: Originally the tower, latterly the prison, which was often the lowest part of the tower keep.
Embrasure: See crenellation.
Espringal: Siege engine.
Garderobe: Latrine. The word is virtually the same as wardrobe (compare guard and ward) and is said to owe its name to the fact that the atmosphere made the place mothproof and hence suitable for storing clothes.
Hoarding: See brattice.
Invest: Lay siege to.
Keep: Tower or main part of castle during eleventh and twelfth centuries. *Shell-keep* was a wall built around the mound as at Berkeley (Glos.).
Machicolation: Stone platform built out from battlements above gateways. It was a stone version of a hoarding. This device was used less by the English than it was by other nationalities.
Mangonel: Siege engine which slung stones, and from which the word 'gun' is derived. It worked by torsion. It was also called a Martinet; however, the present meaning of this word derives from the fact that Martinet was the name of a renowned drill-master in the reign of Louis XIV.

Mark: Coin worth 13s. 4d.

Moulon: Mangonel.

Malvoisin: Lit. Mal Voisin = bad neighbour. The term was first used in 1095 when William Rufus besieged Robert Mowbray in Bamburgh Castle. The malvoisin was then a motte and bailey structure, but the term was soon applied to movable towers.

Millimete gun: The earliest gun of which we have a picture. It is shown on the Millimete MSS. at Christ Church, Oxford, and is dated 1327.

Mine: Tunnel under wall or building. There is a well-preserved example at St Andrews Castle, Scotland; although this dates from 1546 it is representative of mediaeval mines.

Onager: Siege engine.

Perrier: Stone throwing engine; also called a 'bricole'.

Petrary, petraria: See perrier and trebuchet. There was one siege engine to every hundred men (approximately).

Pyx: Container in which consecrated bread was kept.

Sapper: Miner.

Scorpion: Early form of crossbow siege engine; worked by tension.

Slight: (Verb) to destroy defences. Cromwell 'slighted' Wallingford and other castles after the Civil War.

Solar: Originally, raised floor, dais, or gallery in hall. It is said that when William I's sons, the future Henry I and William Rufus, were playing dice on the solar at the Castle de l'Aigle in 1087 they made a lot of noise, and poured water over their elder brother, Duke Robert of Normandy, when he objected. The quarrel, stopped by William I, was said to be the origin of a lifelong feud between Robert and the two others.

Testudo: See tortoise.

Tortoise: Covering for battering ram or miners.

Trebuchet: Siege engine which worked by counterpoise. It was the only siege engine which originated in the Middle Ages.

Notes on Sources and Authors

ORDERICUS VITALIS

Ordericus Vitalis was born in Shropshire in 1075, but after receiving the beginnings of his education in Shrewsbury he crossed to Normandy, where he became a monk in the Abbey of St Evroult in 1085. In 1086 he relinquished his English name Orderic (Latinized to Ordericus) for that of Vitalis, one of the companions of St Maurice. In 1107 he became a priest. He died in 1141 or 1142.

Although the main part of his life was spent in monastic seclusion he was able to recount contemporary events with considerable force and detail. Unfortunately his aim was to write a miscellany of history dating from the birth of Christ, and in doing so he had no inhibitions about chronological order, or even repetition. However, for the reader who is prepared to tackle one of the most discursive histories ever written, Ordericus is not only a mine of valuable information but also a most entertaining and vivid narrator. Although living in Normandy he describes himself as an English monk: and this attitude enables him to view the Normans with less enthusiasm than the purely Norman chroniclers did. Surprise is sometimes expressed at the ability of monks to write about contemporary events, but it should be remembered that monasteries were points of call for travellers, and an alert chronicler would hear numerous stories which he would be able to check and assess against one another.

WILLIAM OF MALMESBURY

William of Malmesbury was born in 1095 or 1096 and died between 1142 and 1150. Little is known of his life except that he spent most of it in Malmesbury Abbey, and that he occasionally travelled to other monasteries to consult and collect books. His history is much more coherent than Ordericus, although not perhaps as vivid when describing stirring events.

RICHARD OF DEVIZES

Richard of Devizes was a monk of St Swithun's, Winchester, in the twelfth century. He was a man of strong prejudices on almost every subject, despised the French, and thought little of some of his English contemporaries. His writing is vivid and informative and is enriched by a lively sense of humour. Devizes was probably his birthplace.

MATTHEW PARIS

Very little is known of the life of Matthew Paris. He was a monk of St Albans and under the name of *Chronica Majora* collected together the works of known and unknown mediaeval historians, to whose writings he added some of his own. Thus the writings of William of Newburgh and Roger of Wendover appear in the *Chronica Majora.*

G. T. CLARK

G. T. Clark wrote a series of articles for various journals, including *The Builder*. These articles were eventually collected and published in 1884 in two volumes as *Mediaeval Military Architecture*. Publication was at the request of Professor E. A. Freeman, of Cambridge. Both these archaeologists have been heavily and scornfully criticized by those who followed on account of certain assumptions they made in ignorance. However, it is easier to criticize than to pioneer, easier to pontificate about battles than to fight them, and it is as easy to find flaws in the work of Clark's critics as they found in his. Furthermore, without Clark's work much of modern deduction could not have taken place. He made careful surveys of many castles and mounds which have since been damaged or altered, and although his historical deductions may be open to question there is no doubt as to the soundness of his architectural accounts.

After leaving Charterhouse Clark became a Civil Engineer and worked on the Great Western Railway. He extended the line to Paddington and built the bridges at Moulsford and Basildon. Subsequently he went out to India where he was Governmental Engineering Adviser. In 1852 he returned to England and devoted himself to the Dowlais iron works, which he made extremely profitable.

Although he spent his last years in Wales (he died at the age of 88) he was criticized in that country for not speaking Welsh, and not praising Welsh achievements sufficiently.

GESTA STEPHANI

Gesta Stephani is an anonymous manuscript. Recently another text of the old MSS. came from Valenciennes where it had lain unrecognized in an abbey library for many years. The result is that we now have a fairly full version of the events and opinions of the reign of Stephen. The author is a strong supporter of Stephen, and scornfully refers to Queen Matilda as 'The Countess of Anjou'.

HENRY OF HUNTINGDON

Henry of Huntingdon was born between 1080 and 1090 in or near Lincoln. From childhood to early manhood he lived with the family of the Bishop of London, who was much concerned with civil affairs. The training he received at this stage, combined with a natural aptitude for business matters, gave him opportunities to observe and comment on events which were beyond the scope of contemporary writers. The date of his death is unknown.

E. VIOLLET-LE-DUC

Colonel E. Viollet-Le-Duc was an architect and military engineer. He advised the Emperor Napoleon III about the defences of France, but his recommendations were not fully implemented. He served with distinction in the defence of Paris.

He was brilliant at deducing military facts about long-forgotten wars. Much restoration of ancient fortresses was completed under his guidance; the work was often done so well that it attracted the criticism of being better than the original, particularly at Carcassonne.

Select Bibliography

Appleby, J. T. (Tr.), *The Chronicle of Richard of Devizes*, Nelson.
Baker, T., *The Normans*, Cassell, 1966.
Barbour, J., *The Bruce of Bannockburn*, Mackay.
Boutell, C., *Arms and Armour*, Reeves and Turner, 1893.
Butler, D., *1066*, Blond.
Edwards, J. Goronwy, *Edward I's Castle Building in Wales*, G. Cumberledge 1944.
Freeman, E. A., *The Reign of William Rufus*, Oxford, 1882.
Hewitt, J., *Ancient Armour and Weapons*, Osford, 1860.
Hewitt, H. J., *The Organization of War under Edward III*, Manchester, 1966.
Howlett, R. (Ed.), *Chronicles of the Reigns of Stephen, Henry II, and Richard I*, Rolls Series, 1874.
Jacob, E. F., *The Fifteenth Century*, Oxford, 1961.
Keen, M., *The Outlaw in Medieval Legend*, Routledge and Kegan Paul, 1961.
Macaulay, G. (Ed.), *The Chronicles of Froissart*, Macmillan, 1895.
McKisack, M., *The Fourteenth Century*, Oxford, 1959.
Morris, J. E., *The Welsh Wars of Edward the First*, Oxford.
Oman, C., *History of the Art of War in the Middle Ages*, Burt Franklin, 1924.
O'Neill, B. H. St J., *Castles and Cannon*, Oxford, 1960.
Paisley, C., *Practical Operations of a Siege*, P.F.P., 1843.
Paris, M., *Chronica Majora* ed H. R. Luard, Rolls Series, 1874.
Partington, J. R., *History of Greek Fire and Gunpowder*, Heffer, 1960.
Poole, A. L., *From Domesday Book to Magna Carta*, Oxford, 1955.
Potter, K. R. (Tr.), *Gesta Stephani*, Nelson.
Potter, K. R. (Tr.), *Historia Novella*, Nelson.
Powicke, M., *The Thirteenth Century*, Oxford, 1953.
Powicke, M., *Ways of Medieval Life and Thought*, Odhams.
Sayles, G. O., *The Medieval Foundations of England*, Methuen, 1960.
Stenton, F. M., *William the Conqueror*, London, 1908.
Storey, R. L., *The End of the House of Lancaster*, Barrie and Rockcliff, 1966.
Viollet-Le-Duc, E., *Annals of a Fortress*, Sampson Low, 1875.
Treece, H. and Oakeshott, E., *Fighting Men*, Brockhampton, 1963.
Watson, J., *Memoirs of the Ancient Earls of Warren and Surrey*, Warrington, 1782.
Whitelock, D. (Ed.), *Anglo-Saxon Chronicle*, Eyre and Spottiswode, 1961.

As the reader will readily observe the above books vary considerably in the quality and authenticity of their information. Sources which have previously been discussed are not included in the above list.

Index